# THE NUCLEAR

# ROCKET

## MAKING OUR PLANET GREEN, PEACEFUL AND PROSPEROUS

We acknowledge the financial support of the Government of Canada through the Book Publishing Industry Development Program for our publishing activities.

Published by Apogee Books an imprint of Collector's Guide Publishing Inc., Box 62034, Burlington, Ontario, Canada, L7R 4K2, http://www.apogeebooks.com

Printed and bound in Canada

The Nuclear Rocket by James Dewar with Robert Bussard
ISBN 9781-894959-99-5  - ISSN 1496-6921

# The
# Nuclear Rocket

## Making Our Planet Green,
## Peaceful and Prosperous

by

James Dewar
with
Robert Bussard

An Apogee Books Publication

In memory of Violet, Louis
and Richard Kaster

# Table of Contents

# PREFACE

I finished my history of the nuclear rocket program (*To the End of the Solar System: The Story of the Nuclear Rocket*, University Press of Kentucky, 2003; reprinted by Apogee Books, 2007) with two chapters that analyzed its significance. I covered a lot of ground there. For example, chapter 16 considered the scientific and economic benefits of spending $1.4 billion on Project Rover/NERVA over an eighteen-year period (1955-1973). Chapter 17 investigated it from many different perspectives, each of which had something to contribute, but which were unsatisfying because they just nibbled around the edges of the nuclear rocket's significance. In the final pages of that chapter, I cited ideas or themes I considered the most important for the nuclear rocket, its vital core so to speak, but concluded they needed at least another book to consider. This is that book.

In beginning it, however, I must sift through those topics and perspectives to clarify for the reader the ones most important from those less so. Telling the story of Project Timberwind, the classified nuclear rocket program for President Ronald Reagan's Strategic Defense Initiative of the 1980s, or of the effort to establish a NERVA-like program as part of President George H.W. Bush's Space Exploration Initiative of 1991, would certainly be interesting and informative, but they would have little impact on the issues I feel are most important. I will say the same for President George W. Bush, who gave the National Aeronautics and Space Administration (NASA) a new direction, part of which saw the space agency reestablishing a nuclear rocket program. But I will not discuss Project Prometheus other than to say it is defunct. And to those looking for a justification for a manned Mars mission using a nuclear rocket - this a return to Vice President Spiro Agnew's Space Task Group of 1969 – I'm sorry, but I will disappoint. You will find no such justification here. To those who may feel inclined to stop reading at this point, I urge patience. I hope this book gives you a different perspective on nuclear rockets and on the space program they alone permit. I think you will be challenged to adopt new thinking, or at least I hope so.

Let me continue my sifting of themes. In chapter 16 of *To the End of the Solar System* I analyzed the $1.4 billion spent on the nuclear rocket and found it produced gains still reverberating through the U.S. economic and scientific sectors. In all honesty, I seriously understated the extent of the graphite and heat pipe industries, which the program helped create, but I see no need to list the firms and itemize their multibillion dollars in annual sales. Instead, I'll just mention graphite golf clubs, tennis rackets or fishing rods derived from Rover/NERVA, as did heat pipes that are used extensively today in laptop computers, satellites or other products requiring the maintenance of an even temperature. Then in Table 16.1, I listed firms that derived from Rover/NERVA. In my research, I called the local telephone operator as well as the Chamber of Commerce to try to locate them, but to no avail. I hoped after my book was published, it might prompt people to contact me with further information. I am happy to say that John Swanson, whose firm was listed as Swanson Analysis Systems, did contact me and said it still exists but is now named ANSYS, Inc. It is listed on NASDAQ and has a market cap of over three billion. Moreover, I did not trace down the patents and discoveries filed by the program's participants, and though that is a fruitful area for research, I will not do it here. I could expand on other scientific and economic successes,[1] such as mastery of liquid hydrogen (LH2), to complete the record, but I wish to look forward, to offer new thinking, and so will conclude my sifting with two comments.

First, in chapter 16 I mentioned the use of Rover/NERVA technology to develop *compact*, high temperature gas core reactors (HTGR). HTGR technology has been around since the 1950s, with prototype reactors built and operated in the United States and overseas, yet these had large core dimensions, and never overcame the light water reactor for electrical energy production. But this might. Its extremely small core size, perhaps the size of a 55-gallon drum, can dramatically lower the capital costs of construction while at the same time making it easier to secure. This should not be surprising, as small building projects are less costly than large ones, and tiny targets are easier and less expensive to secure than bigger ones. Moreover, since this reactor could operate around 1200° C – blast furnace temperatures about four times higher than a light water reactor - thus giving much higher efficiencies, it opens up possibilities for process heat applications. That could revitalize the oil, coal, petrochemical, chemical and other industries that require high temperature process heat by giving them a less costly alternative to electricity, oil, coal, or gas.

Perhaps most important, however, this type of reactor could be decidedly Green, an environmental asset, by creating a totally new industry to process the wastes that are clogging up our landfills and slowly contaminating our groundwater. Up to now, the only real solution for such wastes has been to bury or burn them, with the former the most widely used. However, this blast furnace temperature could open up a third option, *melting* the wastes into a liquid that then is separated through chemical, mechanical or other processes into the useful and thus commercially valuable and the true waste, the latter now in concentrated form for easier, perhaps relatively risk-free disposal. No more leaching of contaminants to our groundwater. If this proves workable, it could allow the creation of centralized processing areas, perhaps in underemployed or Rust Belt areas.

And industries should spring up around these centers to process the valuable liquids into commercial products, bringing even more jobs to those areas. Thus, if feasible, it could mark a real step forward to the Green goal of sustainable development that recovers, refines and reuses as much as possible of the wastes of human civilization, leaving only the bad and nasty behind and even that, as I will show in this book, might be disposed of economically and permanently. In sum, these possibilities are not a NASA activity, but properly belong to the Department of Energy (DOE) and Environmental Protection Agency (EPA), perhaps in coordination with the Department of Commerce (DOC). They should not be rejected out of hand, for fear of terrorism or an emotional anti-nuclear bias, for example, but be subjected to feasibility studies. If those prove promising, then a normal research and development sequence should start and end, if all goes well, with pilot projects to demonstrate the technology to the private sector.

Second, I suspect such uses will prove technically and economically viable on Earth; if so, they will be even more viable in Space because ferrying supplies for lunar bases and planetary outposts would be expensive and dangerous. Compact, high temperature gas core reactors could provide not only electrical power, but also heat to process the soil, rocks, and atmosphere (if there is one) into commodities vital for the astronaut's/settler's survival, thereby lowering the danger and cost of such bases. To rely on solar cells is a perilous folly. After the bases gain a foothold, such reactors might process commodities that have value back on Earth, making the settlements for-profit enterprises. I admit this vision is quite remote, even far-fetched now, when the Space efforts of most nations are centered on keeping a finger-hold on the space station. However, the vision is inherent in the technology of using the energy within the atom to propel rockets.

Let me now sift through the ideas and themes I covered in chapter 17 of *To the End of the Solar System*. I began with nine recommendations for reconstituting a nuclear rocket program. Some are important for this book, so I will discuss them. I followed this with legislative-executive branch analyses – the traditional way to view government programs. Though meritorious and academically interesting, they fail to grasp Rover/NERVA's importance. I will not discuss them here. Then I debunked the "lack of mission" and "ahead of its time" explanations for the program's termination. This book, I hope, not only further discredits, but also totally eliminates that way of thinking.

Next, I proposed that the real credit for the space program of the 1960s belongs not to President John F. Kennedy but to a cabal (a word with sinister connotations but which may not be off the mark). This group, which included Vice President Lyndon B. Johnson and others in Congress (including Senators Robert Kerr, the "King of the Senate," and Clinton P. Anderson) took a space program that received one percent of the budget in 1960 to five percent in 1965. At that time the federal budget was $100 billion. Hopefully, I provoked debate with those who adhere to the Kennedy myth, but that topic is unimportant here. The pivotal role of the budgeters, the Bureau of the Budget/Office of Management and Budget (BOB/OMB) in the space program of the 1950s, 1960s, 1970s and even today, is self-evident. In this book, I argue a properly constituted nuclear rocket program would lead to a dramatic transition from public to private funding of the space program. This would leave them with a vastly different and more restricted role.

Finally, theories of history have captivated scholars and the public for centuries, yet some fail to account for the appearance of Fortune, the unexpected and unforeseen. When that happens, the theory suffers. I placed Walter A. McDougall's "technocratic state" theory (in his book "...the Heavens and the Earth) in that category and see no reason to change my view. I will not consider theories of history further here.

Also in chapter 17 I stated the nuclear and space programs have been the primary drivers of the U.S. economy since World War II (WWII). One reviewer of To the End of the Solar System dismissed it as "overzealous boosterism" and faulted me for not quantifying it (www.thespacereview .com, April 21, 2008). I'll try here because this economic surge will form an important backdrop to my argument, and I will argue a properly reconstituted nuclear rocket program can have a similar one, but this surge will last a long, long time.

Originally, I made the statement to challenge a Brookings Institution study (Atomic Audit, 1998) that reviewed the $5.5 trillion costs of the U.S. nuclear weapons program (very broadly defined) from WWII through 1996, and implied the money was spent unwisely – in other words, a "rat hole" – and detracted from economic growth. Part of this "rat-hole" spending, Brookings held, was the $1.4 billion spent on the nuclear rocket (I've already rebutted that assertion). Yet, it listed things such as computers, nuclear power plants, machine tools, the Internet and Global Positioning Satellites (GPS) and, most ironically, the Brookings scholars said they "do appear to have valid uses in a peace-time economy." Appear? That verb is the study's fatal flaw. Let me begin by quantifying some things that derived from this $5.5 trillion and trace how they have impacted our economy. I won't focus on the atomic bomb effort during WWII, but start in 1950 when the Pentagon spurred development of the cryogenic liquids for rockets, as by then nuclear weapons were becoming small and light enough to be carried by missiles.

Cryogenics: Before 1950, firms

such as Air Products and Linde sold compressed gas for welding, but by 1960 produced cryogenics like liquid hydrogen, oxygen and nitrogen in economic, tanker-size quantities. With military needs met, the firms now had excess capacity; soon the use of cryogenics raced through the steel, chemical, food processing, and other industries, and firms like Air Products were on their way to becoming multibillion dollar international giants.

*Computers*: The nuclear weapons program emphasized their advance, starting with John von Neumann's MANIAC, and going from there. Today, our personal, business, and governmental lives are dominated by the computer – it doesn't just "appear" to have uses in a peacetime economy, it *runs* economies worth many more trillions than $5.5.

*Nuclear power plants*: The light water reactor dominates the world in producing electricity, and its lineage goes back to Admiral Rickover and the nuclear navy. We get 20 percent of our electricity from nuclear and France and Japan 80 percent.

*Machine tools*: In WWII, our machine tools had accuracies to 1/10,000 of an inch; by the 1980s, the last time I followed it, we had accuracies to one-millionth of an inch at the little known Y-12 site.

*The Internet:* Most everyone knows the Defense Advanced Research Projects Agency invented it,

but few know why. Here's why in a nutshell: When Eisenhower became president in 1953, our arsenal was several hundred fission bombs, but our foreign and defense policies became that of "massive retaliation" and "a bigger bang for the buck." That caused an enormous buildup in nuclear weapons, now including H-bombs. Back then nuclear weapons were tested to further their development as well as to see their effects. In 1958, the Teak, Orange, and Project Argus shots occurred at ever-higher elevations, creating a lot of disruptive electromagnetic pulse, but in July 1962, the Starfish Prime (a 1.4 megaton shot 250 miles up in the South Pacific) knocked out communications in Hawaii and destroyed satellites. Then in October 1962, the Cuban missile crisis occurred with the Soviet Union. This caused a loud "Whoa Nellie!" as the Pentagon realized an exchange (a euphemism for nuclear war) could blackout communications. Well, at that time and to paraphrase Soviet Premier Nikita Khrushchev, we were turning out nukes like sausages; our arsenal was peaking at over 30,000 weapons and 75,000 megatons. Some of these lacked PALS (devices to prevent their unauthorized use). So the problem became how to communicate with our forces if an exchange happened to ensure a silo sitter, sub captain, dashing fighter pilot, or grizzled sergeant didn't operate with a "use it or lose it" attitude and make a grisly situation worse. An answer became a vast network of computers linked to each other throughout the country in such a way that if some were destroyed in an exchange, those remaining could still "talk" to each other. This became what we now know as the Internet.

*Global Positioning Satellites*: It originated from the Pentagon's desire to avoid collateral damage in an exchange with pinpoint accuracy and thus minimize the kilotons/megatons needed to take out a military target. GPS is ubiquitous now, in airplanes, boats, cars, and even cell phones.

Moreover, in my government career, I worked extensively with the various facilities comprising the nuclear weapons complex, and learned while they fabricated all the bombs for the stockpile, this complex essentially produced only plutonium, highly enriched uranium (HEU), and tritium. Nearly everything else – the high explosives, the machine tools, the diagnostics equipment, the computers as well as the people with the skills, knowledge, and expertise to use them – came from the private sector. The means for getting things through the barrage of guards, gates, and guns at these government facilities was the contract and the competitive bidding process, the latter being large in scope as it often involved assisting firms to upgrade their capabilities to produce the items to the required specifications. The means of getting the skilled men and women through the gates was the employment contract, and just because they now had security clearances and an obligation to abide by the rules of classification and security, that did not

eliminate their constitutional rights of freedom of speech and assembly. Indeed, these skilled workers had not only the right but also often the obligation to interact with the private sector, particularly when working to establish the specifications for items; and they had the right to quit at any time and take their skills to another employer in the private sector.

I further learned how extensive and pervasive the interaction was between the weapons complex and the private sector when, in the late 1980s, I was part of a group that added technology transfer to its mission. It was stunning, even where one would expect few contributions such as the Nevada Test Site, for example, in precision drilling, diagnostics (the multibillion dollar firm of EG&G started here), and environmental monitoring, including the "Atomic Ranch." I could continue with the nuclear weapons laboratories or facilities such as Los Alamos, Livermore and Sandia, Bendix and Mound, Pinellas, Y-12, Rocky Flats, Savannah River and Hanford. Then to my overall theme I would add the DOE civilian laboratories such as Argonne, Brookhaven, and Oak Ridge.

Space does not allow me to continue my rebuttal, and I've grievously overlooked NASA and its laboratories as well as those in the DOD. The same goes for private industry, those companies that supply goods and services for the nuclear and space programs, and for groups that study the weather and water. It may shock the reader but they were deeply involved in nuclear weapons development and operations, and the tools and skills they developed are now used in the civil sector. Finally, I have room for only a passing reference to disciplines from archaeology to zoology that use products or services deriving from the nuclear and space programs. Carbon-14 dating, anyone?

This summary should answer the "overzealous boosterism" charge as well as refute the Brookings Institution study. So I come back to my point: The nuclear and space programs have been the primary drivers of the U.S. economy since WWII. Brookings was right to conduct an audit, but its scope was entirely wrong. It should have asked why the United States prospered so mightily while the Soviet Union did not. It's a classic for analysis: Both countries started their programs at the same time, with the highest national priority and their best people, and on the same physical terms: Gravity affects all rockets equally and plutonium and uranium have the same properties whether they are U.S. or Soviet made. So why did one produce a cryogenic industry, computers, the Internet, and GPS and not the other? The Soviet Union had a nuclear rocket program; why didn't it develop graphite and heat pipes and master LH2? Why did one succumb to the Strategic Defense Initiative (an implementation of what was learned from Starfish Prime, et.al.) and not the other? One answer is that one had a decentralized federal government system with a Bill of Rights granting its citizens freedom of assembly and speech, and the other did not.

Not merely a dry scholarly issue, this has policy implications. Will KGB Russia, flush with oil cash, be a long-term threat as long as Moscow keeps its tight control? Or will it be only a short-term nuisance, one destined for the scrap heap of history like its Soviet Union predecessor when the oil money runs out? That, however, is not important for this book, but another, related topic will be. The U.S. taxpayer funded this mighty surge, but John Q. Public got little direct return from it. He or she only received an indirect return – more jobs and products and so forth - a trickle-down effect. That's the best that could have been obtained given the times. However, I hold a properly reconstituted nuclear rocket program would have a dramatic economic stimulus, over time even more stunning than the surge of the last half century, this time a surge lasting a millennium. Those are truly audacious words, and in my argument I will show how and why it can happen and seek to structure it so individual citizens can participate and prosper from it. To put it simply, I want the "great unwashed" to cash in for centuries, and not just companies or individuals in the know, as in the previous fifty years. This taxpayer's return on investment attracted considerable attention in the 1950s but has disappeared, regrettably, as an issue today.

This concludes the sifting. I will develop these topics in greater detail in chapter 1 as well as outline the approach for my argument. Moreover, before the reader begins, I'd like to clarify one term he or she will see frequently, that is, the term megawatt or MW, and it means 50-pounds of thrust when applied to rockets. Thus, a 1500 MW engine would have 75,000-pounds of thrust, a 3000 MW one 150,000-pounds, and so forth.

Finally, I am the author of this book, and while I wrote the sentences and paragraphs contained herein, Dr. Robert W. Bussard had a key role in guiding what they say. Those who read *To the End of the Solar System* will know he was instrumental in starting the nuclear rocket program in the 1950s, and maintained an active and professional interest in it all his life. Bob helped me develop my interest in nuclear rockets as a graduate student years ago and reviewed manuscripts for this book, making suggestions, some of which changed my argument.

I am deeply indebted to Bob's contributions and consequently list his name in the credit to this book.

I also have been fortunate to have a number of people review and comment on the manuscript, and I wish to thank them for their efforts. This list includes Rick Ballard, Gary Bennett, Frank Durham, Harold B. Finger, Norm Gerstein, Milton Klein, David Livingston, Eugene Robinson, and of course Stan Gunn, who, as always, was particularly helpful. Both he and John Napier helped write Appendix B on the development of first- through-fourth-generation engines. I also pay special thanks to Ted Spitzmiller, a fellow author at Apogee, who reviewed the manuscript and helped write Appendix C on the flight profile. However, I alone am responsible for the views contained herein.

James A. Dewar
Oxford, Maryland

# Introduction

This is an amazing, strange, remarkable and wonderful book. It is a visionary exposition of a hopeful and imagined future where world peace is achieved, and human cooperation assured by the forces of industry, space flight and commerce, driven by the economic benefits for all that these things offer.

It is not really about nuclear rocket propulsion or its applications to space flight, though it certainly discusses such, but it is about the consequences of nuclear rocket propulsion on the nature of societies, governments and people across our closed planet. It is a book about what might be made to happen if we can organize ourselves to do things right.

And that, of course, is the big if.

Its author, Jim Dewar, is a truly visionary man – and he has had these thoughts for a long time – ever since he lay on a grassy hill one night and realized that the stars were our true home. But his ideas are not some utopian scheme, a favorite subject of philosophers, but are tied to a pragmatic realism, as he could see ways to make them happen in the real world. His experiences in government have not stifled him, but rather, given him insights into the workings of the world that are used in the theses in this book.

I have known him for a very long time, too, since the early days of the Rover program, and we have had many happy and informative hours together over the years, trying to think through what to do and how to do it to get us off this planet in an effective way, one not tied to government spending. That certainly is ineffective, linked as it is to the annual budget cycle and the government-wide competition for funds. Jim has taken the additional steps of seeing what this could mean to the future of our Earth-bound societies and people in terms of their cooperation or conflict and in terms of long-term human growth and mutual progress if private sector funds were involved.

Expansion into Space – in a practical, economic way – is the keynote of the whole thing, and this book is based on his conviction that there is actually a way to do it now, and I emphasize 'now' as well as I emphasize economically, based on the half-century old technology of the nuclear rocket program of the 1950s and 1960s. He is right, of course, but the US seems to overlook all this in the present turmoil of global strife – wars, the environment, energy shortfalls, everyday politics and all the other issues that seem to occupy all our governments.

Jim's conviction of a vision of world peace is stimulated by Leo Szilard who said in 1932 the discovery of nuclear energy could open the door to space flight and this would bring mankind together for a peaceful new expansive and adventurous enterprise for all people. On the face of it, this seems silly but the young Hungarian refugee had a knack of being right. One of the young Turks of atomic energy during the 1930s, he foresaw the industrial uses of atomic energy, something Ernest Rutherford, the most famous of the atomic scientists of the first third of the twentieth century, declared to be only "Moonshine." Well, with over four hundred nuclear power plants operating world wide, I would say Szilard was quite right. He also influenced another refugee – Albert Einstein – to sign a letter he drafted to President Franklin D. Roosevelt that led to the atomic bomb effort during World War II. His instinct here on the military applications of atomic energy was quite right also.

Many others since have held similar hopeful views of the role of human expansion into Space, that it could and would stabilize us against conflict, bringing us together in a shared enterprise of the greatest import, excitement and magnitude.

While these noble and uplifting thoughts are meritorious, I do not share these views if it means the peace of Isaiah, where lambs lie down with lions - in other words, no human conflict, no violence, and no war. And I note Jim doesn't either. The history of human progress and social evolution does not offer much support for this type of peace either. Robert Ardrey's seminal book The Territorial Imperative (1966) shows clearly what and why we are here.

We have gotten where we are because we are aggressive and combative and have triumphed over all the many obstacles that faced us over the long course of millions of years of evolution. We became creatures of large brains, able to think through how best to defend ourselves and attack others, all to our eventual benefit. This is all built into our DNA by thousands of millennia, as so eloquently described in Matt Ridley's book Genome (1999) showing the progress of the selfish gene that has made us so successful as a surviving species. Indeed, my own experience tells me that this aggression-seeking internal intellectual drive is active when it occurred to me that the very means I had conceived for interstellar flight (the ramjet) could be used to blow up stars!

Or, as George Bernard Shaw wrote in Man and Superman "If you want to see Man's greed and sloth, look at his factories, belching smoke and ashes; but if you want to see his beauty and intellect, look at his weapons." This point is well illustrated by Ardrey who observes that 95% of the early anthropological "tools" in the British Museum are, in fact, weapons.

Stanley Kubrick brilliantly encapsulated this in one of the early scenes in his classic movie "2001" where the pre-human ape discovered a large bone as a crushing weapon and throws it into the air in delight, which – spinning – transforms into a beautiful, large orbiting space station. We have gotten to where we are because we are warlike and aggressive. It is not likely that this inherent trait will disappear – certainly not soon.

However, this reality in no way invalidates the vision, as many types of peace exist other than the one of lambs and lions lying down together, and are certainly worth striving for – we cannot continue to destroy and defeat ourselves now that the planet is a closed finite-dimensional sphere on which we are all 25 minutes from each other by ICBMs and nuclear weapons. We have discovered too much for our own safety, unless we find a way to control our inner drives and use them to disperse ourselves off this fixed small place.

Perhaps, just perhaps, if we do so, we might come to enjoy each other's company, especially if we have found a lot of room in which to expand and be free. And that is what practical space flight really offers – the chance to go forth, seek and be free again. A start is made into the solar system, and once that is achieved, then comes star flight. But we must start, and that is what we almost did in the last century with the Rover program, but then failed to continue it in favor of our drives for aggression.

Jim's vision of world peace through human expansion into the solar system puts flesh, muscle and bone into Szilard's insight and drives this book. It is a valid yearning and something like this might well occur if all things could be made to go as he described. So, indeed, Szilard might be proved right here also.

And that is exactly why this book is so important – it is the first outline of a way to proceed that might really succeed. It is a most odd book in that it strives for a visionary hopeful world, but Jim has shown how this might be attained by very non-visionary practical and pragmatic means based on actual engineering results from work a half century old. This is something that has never been put forth in this way before. The space program has never had a long-term view of the sort described here. And it never will, unless it can start down the roads he has outlined.

What he has seen is the simple fact, so often overlooked, that the engine performance from chemical rockets forever limits us to earth or to just tinkering several hundred miles up, just fiddling really at great cost, most of which is born by the taxpayer. In other words, an "energy barrier" exists that chemical rockets can never, ever overcome. This barrier is very real, and is a result of the basic physics of space flight and planetary gravity fields. While we can attack each other across the planet with chemical rockets, we cannot achieve practical space flight for the human race without going to much higher energy levels than are possible by burning fuels.

This is just a simple fact; the gravity well of planet earth is too great for the energy of the chemical bond to give us rockets of practical size and scale and cost to allow our expansion into space to take place. We did send a few men to the Moon, but the cost, size and scale to do so was gigantic – a naval cruiser to put two people in a golf cart on the Moon! But we need this gravity well - else we'd have no atmosphere and could not live here. It is a conundrum. We have a gravity well that keeps us alive, but prevents us from leaving, unless we turn to a higher energy source than chemical combustion.

In the last century, we have discovered this new energy source, nuclear fission, which gives us the ability to destroy ourselves with advanced weaponry. But oddly enough, the energy available from the elementary use of this new, potentially destructive energy source allows us to build rocket engines of such performance capabilities that we can leap the energy barrier that stops chemistry. In other words, we have been given the means to escape our trap of aggression and conflict by the same energy source that has made them so terrifying and implacable. It is well past time to use them for this saving purpose.

The Newtonian equations for rocket flight show simply that throwing small amounts of rocket propellant out the back of a space vehicle is more and more useful if the mass ejected is thrown at high speed. In the equations this appears as an exponential term, which means that a little bit of increased propellant speed goes a very long way to improving the size, scale and practicality of rocket vehicles.

And what the Rover program showed is that we can and did improve this ejection speed by factors of about three, which easily pushed the rocket designs into the 'escape the barrier' regime. Nuclear rockets, using Rover technology, are already good enough to make space flight practical for all humankind. I repeat, it was already good enough.

But it was not to be. The early goals of Rover aimed at a manned Mars expedition for the 1978 opposition mission – which was well within reach of the technical developments begun in the mid-1950s. However, political infighting, management decisions, and interagency turf wars delayed the program enough that it missed its political "window of opportunity" in the 1960s and died as the government focused on fighting the Vietnam War instead of going into space. Ardrey's imperative, indeed!

Yet the engine systems were tested, did work and then lay dormant. Jim has gone back to this array of success and picked out the main technical line of the engines to use as the basis of his arguments for development and expansion into space now. This is all very real and very much attainable – all we have to do is do what we have already done, once again. He has not tried to invent new magical means of achieving orbit or traveling to Mars, but just used what we know how to do though we never used it when we had it.

He is not trying to "sell" the KIWI-

B4 engine, or its successful, long-ago technology, but uses it and its design basis to illustrate that there presently exists a way to proceed without new inventions. And further, he shows that new materials and structures found since that time now allow still more increases in performances to factors of four to five times above standard chemical rockets. Starting with what we had, there is no way to go but up! We do not need space tethers, fusion drives, antimatter or the magical imaginations of zero-point energy, vacuum fluctuation or wormholes to make space flight happen. We just need to use what we have built, tested and proven.

But the beauty of this book is that he has found and created organizational, political and economic plans to evolve the onset of practical space flight in stages, driven by both government and the private sector. This will not be another massive government funded Apollo program. In other words, he has found a way to make Space pay for itself, as it goes, in a manner beneficial to investors and the public. He has found a way to bring the public into the enterprise in a voluntary yet participatory manner in which all citizens, everywhere, can have a stake in the expansion and feel it in their bones, hearts and minds.

Jim's years in government have given him a rare and brilliant insight into the workings of the governmental and political process (as shown in his first book To the End of the Solar System) and he has used this insight to good effect in outlining a promising way to proceed. All that is required is that our government open its mind to something beyond the annual budget and four year presidential election cycles and think of a longer future.

While this seems unlikely today with our current preoccupations on energy, oil, politics and war, we can hope for better days. Future governments may have time and interest to consider the longer future of the American dream and culture and should well attempt to find ways to keep our frontier visionary nature alive. The NASA Apollo program of the 1960s certainly captured the hearts and minds of our citizens, in which all of us gloried in the technical, social and psychological triumphs of going to the Moon. Be we did not share in the economic fruits of this enterprise, which Jim has found a way to do.

John F. Kennedy's death ensured the end of such thinking – it was replaced by the Vietnam War planners – but the technology is still there and, I believe, the interest and excitement of all our people is still there too, waiting to be rekindled for the real adventure of human expansion into the solar system. My hope is that this amazing and visionary book can kick start this adventure once again.

If it does, we will have created something similar to what our founding fathers brought forth over two centuries ago, an idealistic vision of the common man governing himself. Kings and queens and czars and emperors were not necessary. The first shots of the American Revolution indeed were heard and still are heard around the world. In like manner, I believe the first launch of a nuclear rocket will be felt irrevocably around the world, so stunning are its consequences to those who inhabit this finite-closed sphere.

We are blessed to have such as exposition of vision and reality and plans and prospects to consider. Read it, enjoy it, think about it, and do something to make it happen.

Thank you, Jim.

Robert W. Bussard
Santa Fe, NM

July 2007

Photo Credit: Dolly Bussard

### Author's Note:

Robert W. Bussard, "the father of nuclear rocketry," passed away in October 2007. For those who may not know, in 1953 Bob wrote the right paper on the nuclear rocket that was read by the right people in the right places at the right time. It led directly to Project Rover/NERVA, and was the first of many papers and books he wrote on using the energy within the atom to propel rockets through the vastness of Space. I'll just mention his Interstellar Ramjet and ASPEN. For the last several decades of his life, he formed a company and led its research team to develop his QED fusion reactor concept. While the latter was nominally for naval propulsion, Bob's real motive was, as always, space flight, as he had a consuming interest since the age of seven to travel to Mars, and as he realized quite early – as a teenager - that chemically propelled rockets couldn't do it in any meaningful or lasting way. Instead, his was a life dedicated to allowing mankind to move off this planet permanently and doing so requires nuclear. I hope his successors make his non-radioactive fusion concept a reality. If that happens, as I hope, we need not consider the solid core nuclear rocket discussed here, yet we must consider much of the argument contained herein. That dialogue must take place and I, along with his many friends, regret that he will not be here to add his expertise and insight as well as civility and humor to the debate. That will be deeply missed, as will his optimistic signoff on his emails: 'Cheers, RWB.'

So Cheers to you, RWB, and thanks for a lifetime of work to bring a new vista for humankind: extending our civilization into the solar system.

# Chapter 1
## Ideas and Themes

In 1960, NASA prohibited the use of nuclear rockets to reach low Earth orbit (LEO). In other words, it banned starting a nuclear engine on the ground or in the upper atmosphere (around 100,000-feet) to propel its payload into LEO. No, it had to start in LEO and go deep into the solar system away from Earth where its "radioactivity" could "do no harm." That decision has had devastating consequences. In this book, I argue repealing this prohibition will allow Congress and like-minded foreign legislators to create a privately funded, prosperous, and democratic space program that leads to world peace and the environmental goal of sustainable development. That's a colorless academic sentence. Let me say it in plain English: Break the taboo and get great political power in the United States and overseas to create a new epoch-changing, fraternal space program with countless new industries and jobs, billions in private investment and a vastly expanded tax base, and with world peace, ultimately managed by citizen investors, and a Green planet as consequences. Well now, that certainly is bold, particularly the world peace and Green planet parts, but can I back any of it up? I think yes. Yes, to every word of it.

To convince the reader of the truth of so audacious an assertion, I obviously must develop my argument logically and rationally, and as it will have many ideas and themes, I must organize and introduce them in an orderly way. I find the best way to start is to restate Leo Szilard's comment I quoted in my first book. It is the seminal idea, from which many but not all of my other ideas and themes derive. In 1932, a friend remarked that mankind had a basic need to fight and would war forever on Earth unless provided an outlet. The Hungarian scientist responded: "The only thing I could say was this: that if I came to the conclusion that this was what mankind needed, if I wanted to contribute something to save mankind, then I would probably go into nuclear physics because only through the liberation of atomic energy could we obtain the means which would enable man not only to leave the Earth but to leave the solar system."[i] Szilard did go into nuclear physics, making truly important contributions, but to my knowledge he never expanded on his amazing and prophetic idea that looks first at the earthly consequences of atomic propulsion – saving mankind from war – and only then at the missions it permits – forays into and then *beyond* (italics added) the solar system. I agree with this emphasis, and it will play an important role in my argument.

This seminal idea seems to have motivated many early proponents of nuclear rockets; in other words, it seems to morph from 1932 into the 1940s, from Szilard to Stan Ulam, a Polish mathematician, who began

**LEO SZILARD**

Photo Credit: Department of Energy

A Hungarian émigré, Leo Szilard was perhaps the most prescient scientist of the twentieth-century. He filed patent applications on critical twentieth-century inventions - the cyclotron, linear accelerator, electron microscope, nuclear chain reaction, and nuclear reactor. With Albert Einstein, perhaps the most famous scientist of the twentieth-century, he collaborated on filing more than forty-five applications to improve home refrigerator design (as they caused many deaths in the 1920s); they sold some patents for tidy sums. In 1939, he wrote the letter to President Roosevelt – Einstein only signed it – that led to the atomic bomb program. Several years earlier, he had expressed a visionary belief in the industrial uses of atomic energy, a conviction contrasting with the most famous scientist of the time, who declared such to be only "Moonshine." Today, his vision has proved correct, as more than four hundred nuclear power plants operate worldwide. After WWII, Szilard sought to bring the arms race under control, and founded the Council for a Livable World, which has played a key role in passing many arms control treaties.

thinking about them in 1944 while working on the atomic bomb at Los Alamos. He certainly knew Szilard but no records exist, so far, to indicate whether they discussed the Hungarian's 1932 remark or whether Ulam developed his interest and insight on his own. My best guess is that Szilard somehow introduced his thought during the war, in the bull session speculation about atomic energy in the tight-knit Los Alamos community - though I admit others, like Ulam, might have had it independently. After all, some of the brightest people in the world were at Los Alamos building the bomb. In the early 1950s, Ulam co-developed the H-bomb with Edward Teller of the Lawrence Livermore National Laboratory (but preferred to let Teller have the spotlight); later he became the nuclear rocket's "godfather" in Washington politics, holding it was more revolutionary than the H-bomb. But he did not say how.

Again, the idea somehow morphs into the 1950s. It may have influenced Senator Clinton P. Anderson (D-N.M.) but I cannot say how or through whom. My best guess is Ulam, as the two were on a first name basis. I discount the possibility Anderson had it on his own. A former insurance salesman, with only a year of college, the senator became chairman of the Joint Committee on Atomic Energy (JCAE) in 1955 and from then on, talked continually with the leading figures in atomic energy, including Ulam. Anderson learned of the nuclear rocket program in 1955, and became its principal political proponent in 1956. That year, well before Sputnik, he said, most remarkably, the colonization of the solar system should be international to avoid wars for empire, and this amazing statement qualifies the seminal idea in a major way. He said colonization, not exploration, of the solar system; in my argument, I will clarify differences between the two words.[ii] Then in 1958, in a speech on the Senate floor, he said nuclear rockets might bring peace to Earth by lifting men's minds from their earthbound hatreds into the solar system. This, of course, paraphrases Szilard's much earlier comment, yet it adds more specificity. But who influenced him to say it remains a mystery; after Sputnik, many space experts roamed Capitol Hill, and they would be likely candidates. Still, my best guess is Ulam.

In a briefing on Project Orion, the nuclear-pulse rocket concept that Ulam conceived at Los Alamos during the war, the celebrated Danish physicist Niels Bohr added still more specificity to the original idea. In 1960, he "saw it (Orion) as a way to combine global nuclear disarmament with a cooperative project with the Soviet Union to explore the solar system."[iii]

**STAN ULAM**

Photo Credit: Los Alamos National Laboratory

In 1944, a year before the explosion of the first atomic bomb, a Polish mathematician, Stan Ulam, conceived the idea of using them to propel a rocket, the bombs exploding sequentially at the base of a rocket, pushing it forward. This nuclear-pulse rocket became Project Orion in 1957, and was studied until 1965 when it was terminated. Ulam became the "godfather" of the Rover/NERVA program in the 1960s, and advocated its development in the White House and Congress.

**CLINTON P. ANDERSON**

Photo Credit: Los Alamos National Laboratory

Clinton P. Anderson (right) (D-N.M.) served in the Senate from 1948 to 1972 and as an "insider," became a powerful advocate of the nuclear rocket program. A month after the Cuban missile crisis of October 1962, he arranged for President Kennedy (left) to visit Los Alamos and the Nevada Test Site, where nuclear rockets were tested, to learn more about the program. Anderson's health declined in his final term and hindered his advocacy for Rover/NERVA, which was terminated officially on the opening day of the new Congress in January 1973. In the background over Kennedy's right shoulder is James T. Ramey, AEC commissioner.

This is government to government, not Anderson's more social vision of men's minds being lifted. Obviously, the Soviet Union no longer exists, and global nuclear disarmament might be a long-term consequence of a nuclear rocket program, when men's minds are indeed so lifted, eliminating the need for not only nuclear but also conventional weapons. However, this prospect is so long-term that I will not discuss it further (indeed, Article VI of the Non-Proliferation Treaty links nuclear disarmament to a treaty on general and complete disarmament). However, inherent in Bohr's idea is using Space to achieve immediate nonproliferation and nuclear disarmament objectives between nations, not some all-encompassing global scheme. That idea is lost today as most states with space programs use them for civil, national security or science purposes. I will argue the nuclear rocket can play a crucial role here.

Other famous figures commented on the world peace theme though they probably never heard of Szilard's idea. In 1965, when faced with a budget cut, James Webb argued the space program would create a universal society, which presumably would be peaceful; though it can be dismissed as a mere budget shenanigan, I think deep down the NASA administrator believed it. And in 1966, arguing about future space goals before Vice President Hubert H. Humphrey, who chaired the Space Council, Glenn T. Seaborg, chairman of the Atomic Energy Commission, urged it be aimed at coloniza-

### NIELS BOHR

Photo Credit: Department of Energy

With an international reputation equal to Albert Einstein's, the Danish scientist Niels Bohr was briefed on Project Orion in 1960, and observed that the nuclear-pulse rocket offered a way to combine global nuclear disarmament with a cooperative program with the Soviet Union to explore the solar system in peace.

### GLENN T. SEABORG AND JAMES E. WEBB

Photo Credit: NASA

Glenn T. Seaborg (left), the Nobel Prize-winning chairman of the Atomic Energy Commission from 1961-1971, was a strong nuclear rocket proponent. James E. Webb (right), NASA administrator from 1961-1968, had a deeply humanitarian view of the potential of the space program to promote peace among peoples, and advocated the nuclear rocket throughout most of his tenure. However, in 1968, he declined to develop it to a flight-rated system because of the chaotic state of NASA's budget as a result of the Vietnam War. The lunar landings were funded, but the follow-on space program, and the nuclear rocket's role in it, was uncertain.

tion of the solar system. That marked the first time the word 'colonization', and the thought contained therein, was raised at very high governmental levels.

This world peace idea still exists, though in a different form. In 1997, the distinguished scientist Freeman Dyson wrote the colonization of Space could alleviate conflicts on Earth by societies' agreeing to migrate to different parts of the solar system.[iv] This conflicts with Anderson's vision of colonization bringing harmony to people on Earth, as it implies that this will never be possible. Still, Dyson's thought contains that germ of the idea common to Szilard and Anderson of using Space to achieve pacifistic goals between peoples, not governments. In truth, the prospect of shipping millions of people throughout the solar system is most impractical at this time; however, with continued research and development on fusion, rocket engines could appear later this century that make this possible. So although he may be proved right, his timing is wrong, as that nuclear engine technology is not there yet.

To those used to a space program conducted with chemically propelled rockets these comments seem bizarre, weird, and out of touch with reality. Yet they are not. The fact none of this has happened has little to do with the veracity of the insights of these distinguished people, but everything to do with the fact that any space program conducted with chemical propulsion will be limited forever. It will always be elitist, expensive, and inequitable, with the benefits of taxpayer money going only to a few. What we have had for the last fifty years is what we will see for the next five hundred or five thousand, a space program aimed only at exploration. In contrast, a space program conducted with nuclear rockets can be egalitarian, inexpensive, and equitable, where all taxpayers can share in the benefits. It will aim at colonization - a permanent, irrevocable, and inevitable human expansion off this planet – or to put it another way, the extension of man's dominion to the end of the solar system.

So to my peace and disarmament theses, I now add economic, egalitarian, and equitable ones, greatly expanding the topics my argument

**FREEMAN DYSON**

Photo Credit: Institute of Advanced Study

One of the most visionary and brilliant scientists of the last third of the twentieth-century, Freeman Dyson worked on Project Orion in the late 1950s, and was deeply disappointed by its cancellation in 1965.

will cover. Let me qualify the latter. I've already noted Rover/NERVA's $1.4 billion produced tangible gains still reverberating in our economy, but it did not directly benefit those who paid the taxes for it in the first place – though some may have gained indirectly in the form of more or better jobs - a trickle-down effect. This is also true for the nuclear and space spending - the primary drivers of our economy since World War II. In my first book, I raised this theme when discussing how, in 1958, Senator Anderson objected to the patent provisions of the bill creating NASA. He wanted the public compensated since tax dollars would create this wealth. I noted it again in the discussions of Rover/NERVA's contributions to NASA's Tech Briefs.[v] That program, still ongoing, is a modest attempt to disseminate NASA-sponsored technology back to the public via a publication and public relations process. Essentially it says "Folks, here are some things you might find interesting and perhaps profitable. It's come one, come all." Finally, I raised it in my preface in discussing the flawed Brookings Institution study. Those references, however, were brief and admittedly indirect. Here I will consider this issue head-on: How can those who pay taxes in the United States share in the benefits when their money produces valuable goods and services? In other words, to put the question colloquially, how can each taxpayer get a "return on investment" on his or her tax dollars? I stress the phrase "each taxpayer," not some vague return to the "public" typified by the shotgun Tech

Briefs approach. I hold it not enough for taxpayers to receive pictures from the Hubble telescope or a Mars Rover; those are interesting and exciting, but definitely not a sufficient "return." And I will further qualify the phrase by saying it is "each taxpayer in the world," not just in the United States.

Finally, I raised a number of other ideas and themes in my first book, particularly in the last three pages of chapter 17, but I will touch on them only indirectly here. Some will require another book to analyze more fully. However, one thought I will consider is that of Robert Seamans (Webb's deputy in NASA in the 1960s) who proposed, in Vice President Spiro Agnew's Space Task Group of 1969, that the space program should be Earth-oriented.[vi] I agree with Seamans (at that time Secretary of the Air Force) but suggest his focus was too narrow. I will argue a properly reconstituted nuclear rocket program can support foreign and domestic policy objectives, including economic, education, energy, and environmental ones. I'll say the latter differently: It can Green the planet, though that will take time and occur in stages. And a theme not included in my first book, but one that I will consider here, is this: Since nuclear rockets permit a space program that is visible to all, that space program must be open and voluntary to all.

In summary, I have outlined many seemingly disjointed ideas and themes: Creating world peace; more revolutionary than the H-bomb; forging a universal society; extending man's dominion to the end of the solar system; achieving nuclear disarmament and nonproliferation objectives; facilitating the achievement of foreign and domestic policy goals, including education, economic, and energy; Greening the planet and realizing the goal of sustainable development; fostering an egalitarian, inexpensive and equitable space program; visible to all, open to all, voluntary to all; and a permanent, irrevocable and inevitable human expansion into the solar system – colonization. You may think: Whew! And double Whew! This list appears quite over the top as well as random and disconnected. Yet all ideas and themes have a unifying thought: Only the nuclear rocket permits the democratization of the space program and the extension of man's dominion to the end of the solar system. In a phrase, it's "democratization and dominion."

I hold it can be achieved and will consider each specific idea and theme in its appropriate place. My argument will have a minimum of technical analysis and jargon, just enough to make my discussion credible and understandable. Any additional clarification, explanation, or justification will be found in boxed texts, footnotes, or appendices. In essence, I write for the average citizen with an interest in the space program, nuclear affairs, Greening our planet, world peace, and other topics included in the phrase "democratization and dominion," and not for rocket scientists, nor for those who want to build outposts on the Moon or go to Mars. I view such goals as utterly unrealistic, as all require government funding, and most governments today have more pressing priorities.

I hope my argument allows all to see the consequences of a properly reconstituted nuclear rocket program that breaks the taboo. Before starting, however, I must summarize my basic propositions (chapter 2) and assumptions (chapter 3) as some may jeer it's "totally absurd" or "utterly preposterous" to believe the nuclear rocket, which never flew a mission, can have such consequences. Inevitable human expansion, a universal society, world peace indeed! For several thousand years people have hoped for the peace of Isaiah where lambs lie down with lions, and philosophers such as Plato have imagined ideal states where people live in harmony. It hasn't happened in all the centuries since, so why are we to presume it can happen via a program that "never got off the ground!" Fair enough! I can understand the reader being skeptical, or more likely totally cynical at this point, but hear me out. Some of the most brilliant people of the twentieth-century – Szilard, Ulam, Bohr, and Dyson – had an insight this can happen. I think they are right and so offer my vision as to how.

## A Note on Radiation

I must speak now about an 800-pound gorilla in the room: Radiation. Perhaps nothing evokes more fear with the public than radiation. We cannot see, hear, feel, taste, or smell it, and this allows some to use the word "radiation" as a scare tactic to make the public (and the space industry and NASA) cringe. Those who know better understand radiation is undetectable by our five senses, but it is easily detectable by our minds, and the instruments we create can measure neutrons, gamma rays and other particles accurately to parts per billion. These people have written countless books and given innumerable speeches, yet the fear remains. I add this note to those voices, and hopefully it may calm the worries of my readers.

Radiation has been, is now, and will be a fact of life. As you read this, radiation from the depths of Space is piercing your body as it pierces through the planet and plunges once again to the depths of Space. Yet you live. Our predecessors lived in a more radioactive world than today, including radiation-filled caves. Yet they lived, obviously or else we would not be here. Today those who walk on some of the most beautiful beaches in the world, who swallow an unwanted gulp of sea-water, who live near breathtaking mountains, or who frequently fly, all have higher exposures than others who don't. Yet they live. All that is from outside our body; inside, in our muscles, tissue, and bones, lay isotopes emitting radioactivity, and most of this comes from the food we eat, water we drink, and air we breathe. Yet we live. Nuclear power plants would be condemned as "leaky" if they exposed their workers to as much radiation as astronauts receive. Yet all live, and most of those who have landed on the Moon are living into ripe old age. In sum, there is no such thing as a radiation-free zone. We are exposed on our outside and inside to cosmic/solar and earthly radiation constantly. We cannot escape these daily doses no matter what we do. Yet we live, and if we are prudent about our lifestyles, we may live into very old age, our exposures notwithstanding. The body can heal itself and indeed, some scientific data suggests it needs radiation in order to thrive and prosper. Finally, we may also live into very old age by using radiation to kill cancer cells or detect other illnesses. So it can help our body heal. No need to fear.

Some, however, seem to accept cosmic/solar and earthly radiation as "good" since it is "natural," while decrying anything else as "bad," since it is "man-made." This is nonsense – radiation is radiation. What makes manmade radiation different is the fact we control when and how it is released. Once released, its behavior is predictable, so we can take countermeasures to prevent or limit its potentially harmful effects or direct them to where they can do good, such as in radiation therapy. Permit an over-the-top football analogy to explain this further. Some critics puff up the penetrating power of neutrons and gamma rays to do unstoppable demon-like harm, as if they are superhuman 500-pound fullbacks, able to bench press 2000-pounds and run 100-yards in five seconds. Obviously then, any team with them in the backfield never needs to pass and always will run and always will win. This also is nonsense. As this team always gives them the ball, but they always, always run to one spot on the line. Realizing this, the wise coach will move all his defensive players to that spot. The fullbacks are stopped; their predictability gives the other team the win – every time.

Though obviously overdrawn, I hope this analogy illustrates the point. All fission and thermonuclear reactions and most fusion concepts release radioactivity, but it is predictable. On Earth, we know air molecules absorb gamma rays and neutrons, the most energetic of the particles, in three-fifths of a mile, so one could stand a mile away from an unshielded test reactor without exposure. This says nuclear rocket testing can be done without harmful or life-threatening incidents. On Earth, we know radiation ricochets off air molecules back to the source (a phenomenon called air-scattering), but around 100,000-feet where the air is scarce and in Space where none exists, it travels in ever-straighter lines. This says where to start nuclear engines. We know radiation is most intense next to the source but decreases as the distance is increased; this is a principle well known to anyone who warms himself quickly next to a fireplace but who remains cold if standing well away from it. This says locate the astronauts at the front of a rocket and the engine at the rear, and let the distance between the two and the intervening structure (e.g., fuel tanks) dissipate and absorb the radiation. In other words, worry much more about protecting the astronauts and payload from cosmic and solar radiation and the van Allen belt; worry much less over that from a nuclear engine. We know that when a reactor is turned off, there is a blip upward in the radiation's intensity, then it decreases on a predictable, logarithmic scale over days, weeks, months, and years – though never back to zero. This implies how to handle Space operations. Just wait; time is on our side.

I could continue – large doses of radiation damage living and nonliving things though again it is predictable – but that would expand this note into a chapter or book. I hope by now the general reader understands a nuclear rocket's radiation is only a challenge for scientists, engineers, and operations personnel, not an insuperable barrier. Its testing and operation can be done without endangering its development and launch personnel, the public, or the world, and hopefully in the very near future average people, not highly trained astronauts, will go on bold ventures with it. What can be predicted can be managed, provided good people are doing the predicting and managing. That is the key point. As a citizen you must either judge or accept the judgments of people you trust that those doing the predicting and managing are doing those tasks well. Here you should listen to the critics, then judge or accept the judgments of people you trust to see whether their arguments are valid or whether they merely seek to inflame your fears or proclaim building and operating nuclear rockets cannot be done. Realize good people can do it.

———

[i] To the End of the Solar System, p. 10. Cf. Spencer R. Weart and Gertrude Weiss Szilard, Leo Szilard: His Version of the Facts (Cambridge, MIT Press, 1978), p. 124.

[ii] In former colonial countries, the word reminds many of its harsh, repressive and involuntary nature. I will show a positive side of the word colonization and how it may bring about world peace.

[iii] www.tbtaylor.com/CHAP3.wtm, "Change of Heart," p. 3, 10/19/91.

[iv] Freeman Dyson, Imagined Worlds (Cambridge: Harvard University Press, 1997), pp. 150, 158.

[v] To the End of the Solar System, pp. 41, 213ff.

[vi] Ibid, p. 182.

# Chapter 2
## Six Propositions

### Proposition 1:
### Technological Epochs

My first proposition is that technology has political, economic, social, intellectual, and religious consequences. This is self-evident considering the technology-induced, epoch-changing shifts in human history, from the Stone to the Bronze to the Iron to the Machine/Industrial ages. These "buzz-word" phrases, coined for popular audiences but too broad for a scholarly one, still have some truth in them, so I will use them. In the transitions there were changes, there were winners and losers, sometimes on a grand scale. For example, the Bronze Age Mycenaean and Minoan civilizations failed to enter the Iron Age that began around 1200 B.C. and totally disappeared. In our era, the start of the Machine/Industrial Age in 1750-1800 led to the growth of large cities, many of which specialized in certain technologies or skills. New York became the financial center; Pittsburgh the steel capital; Chicago the hog butcher to the world. We still live in that age but also have entered what is popularly called the Atomic Age, as illustrated by the title of the official history of the United States Atomic Energy Commission: This New World.[i] With Trinity, the first atomic bomb test in July 1945, everything changed irrevocably: We live in This New World. The nuclear genie cannot be put back into the bottle. In October 1957, we entered the Space Age when the Soviet Union launched Sputnik, the world's first satellite. I will argue that when the Nuclear Age is combined with the Space Age in the form of the nuclear rocket, its consequences can be even more dramatic than all previous epoch-changing shifts. It allows the creation of a fundamentally new international regime and truly opens a new vista for mankind.

### Proposition 2:
### Infrastructures within Epochs

My second proposition is that technologies within an age share common features and require an infrastructure which governs them and in which they develop. This is also self-evident, but to illustrate it further, I will briefly consider four examples from the Machine/Industrial Age: The railroad, the internal combustion engine as used in the automobile, the telephone, and the airplane. In 1750-1800 they did not exist, but around 1900 they did, in rudimentary yet developing ways, and they exist today in more mature forms. Each had a pioneer or group of pioneers credited with inventing the item, though its origins may be traced to earlier periods or other individuals. For the railroad, it is Richard Trevithick (1771-1833); for the internal combustion engine many but perhaps Alphonse Beau de Rochas (1815-93) and N.A. Otto (1832-91) for the four-cycle engine that powers today's automobiles; for the telephone Alexander Graham Bell (1847-1922) and for the airplane Wilbur and Orville Wright (1867-

1912 and 1871-1948 respectively). They demonstrated the feasibility of their technology, setting in motion activities that continue to this day. Capital was raised; private, semi-private or government organizations were created with savvy management teams to guide the enterprise, to oversee teams of skilled and unskilled workers and, of course, to make a profit so that more and more could share in the wealth the technology created. An infrastructure developed to support the invention, often independent of the organization's control, and this ranged from raw materials to maintenance, repairs, insurance, and decommissioning. Then competition evolved to sharpen the item's price and improve its utility in society. No matter their form (monarchic, autocratic, dictatorial, or democratic) governments then promoted, regulated, and taxed the entire operation, and adapted it to their needs, including civil and defense uses. When international involvement was necessary, governments acted by negotiating treaties. Finally, despite initial doubt or resistance, the public ultimately embraced the technology, thus minimizing but not eliminating criticism.

This pattern is easily observable. Different railroads move people and goods quickly from nearly every city in nearly every country. Many telephone services exist in nearly every country. Different automotive companies compete, and innumerable firms exist to repair, fuel, and insure those autos in nearly every country. Finally, anyone can fly anywhere in the world within hours of purchasing a ticket from any number of airlines. Likewise, the pattern can be traced in the many agencies most governments created to promote or regulate these industries or to use their products for governmental purposes. This includes international organizations, treaties, and standards. All this has left a visible mark on the landscape of most countries – a vast network of railroad tracks, highways, telephone lines, and airports, integrating all parts of a country and the world with one another for travel, trade, and communication. These nineteenth-century inventions, as developed in the twentieth-century, have allowed man to establish dominion over the Earth's surface to twenty-miles up in ways that would be utterly unfathomable to those living in the eighteenth-century. We move millions of tons of goods hundreds of miles daily while they would use oxen or draught horses lumbering a few miles a day to haul a ton or two. We drive automobiles hundreds of miles a day while they would trudge on foot to make ten or twenty. We talk with each other instantly from anywhere on the globe whereas their letters would require weeks or months to arrive. We fly safely from continent to continent in hours while they would endure storm-tossed and lengthy sea voyages.

## Proposition 3:
## Limits of the Chemical Rocket Space Age

In 1957 we formally entered the Space Age, and my third proposition is that it is similar to the Machine/Industrial Age, but with one basic difference. In its similarities, the rocket had its pioneers, including but certainly not limited to Robert H. Goddard (1882-1945) in the United States and Konstantin E. Tsiolkovsky (1857-1935) in Russia. After such experts demonstrated the feasibility of chemically propelled rockets, a pattern identical to that of the railroad, automobile, telephone, and airplane developed. Capital was raised, organizations with skilled management and labor teams were formed, an infrastructure developed, and then governments promoted, regulated and used the rocket for civil and military purposes. This process gathered momentum in the 1930s, accelerated during and after WWII, and burst upon the world scene with Sputnik. Since then, an infrastructure matured, with competitive companies and national and multinational governmental organizations, and treaties to govern the rocket. It achieved stunning feats of science, commerce, and exploration: planetary probes, the Hubble telescope, Earth-oriented satellites, the lunar landings and the International Space Station.

These feats, however, obscure the crucial difference of the Space Age. Unlike other technologies of the Machine/Industrial Age that still have a lot of room for development and maturation, the chemical rocket engine – upon which all these feats of the Space Age hinge – quite swiftly reached the height of its technological development.

I will explain why in plain English, hopefully in a manner that does not offend the expert who likes the precision of the rocket equations and who harkens back to Isaac Newton (1642-1727) and his laws of motion, but equally hopefully in a manner that allows the general reader to say "While it is rocket science, it's not really rocket science, but just plain old common sense."

To begin, a rocket moves forward by expelling a hot gas backward through a nozzle. Obviously then, this hot gas is the most important part of a rocket; without it a rocket just sits on a launch pad, so to it we must turn our attention. With chemical engines, burning a fuel with an oxidizer produces this hot, energy-filled gas that cools slightly as it enters the nozzle throat, converting the jumbled, directionless gas into directed kinetic energy that pushes the rocket in the opposite direction as it leaves the nozzle. This is obvious to anyone who sees a rocket blasting upward, trailing smoke and sound, flame and fury. Looking closer at this gas, however, we see that not all gases are equal in energy content. Some burn hotter than others, a principle those with fireplaces know well – hardwoods such as oak burn hotter and longer than softer woods such as pine. Thus, one part of rocket performance is having the gas as hot as possible.

The other part is its weight, or the molecular weight of the fuel and oxidizer, to be precise. Molecular weight is the sum of all atomic weights of all atoms in a molecule. Hereafter, I will use the word weight alone to avoid adding the word "molecular" before it repeatedly. Gases with a lower weight travel faster out the nozzle than those weighing more, in the same way that a man can throw a light baseball faster than a heavy bowling ball. If we substitute heat for the man, the point becomes clear. So rocket performance, in the final analysis, is determined by temperature and weight: The hottest gas with the lowest weight means the best performing rocket. The pertinent part of the rocket equations state $v \propto \sqrt{T/M}$. This means the velocity of the exhaust gas of a rocket engine is proportional to the square root of the temperature of the gas before it enters the nozzle throat over the average molecular weight of the gas. However, we face an impenetrable ceiling with chemical propellants. Many fuels and oxidizers exist and have been used for decades, such as kerosene and oxygen, hydrogen and oxygen, and alcohol and oxygen; but no improvement beyond them exists or will ever exist. Chemical fuels are tapped out. You cannot get more heat out of them, and you cannot make them lighter. Kerosene, hydrogen, alcohol and oxygen weigh what they weigh forever. These are permanent, unchangeable features of chemical engines.

This has consequences. To discuss them in plain English, I will use a simplified definition of the term "specific impulse." [Those who want a precise definition should consult a textbook or the Internet]. It is the best measure of a rocket engine's performance and is popularly defined as the pounds of thrust produced from the pounds of fuel consumed per second. Thinking of it as miles per gallon for an automobile is a helpful though imprecise analogy, as miles per gallon and specific impulse are measures of efficiency. If one auto gets 20-miles per gallon and another 30, the efficiency difference is quite apparent, and everyone knows it will cost more in gas to go 1000-miles with the former than the latter. Rocket engines are similar; those with higher specific impulse will allow the rocket to travel faster and carry more payload than those with lower specific impulses. This is particularly important in sending payloads into orbit because Earth's gravity well must be overcome, and this requires a tremendous amount of fuel. Indeed, 80-90 percent of the weight of a chemically propelled rocket sitting on the launch pad is just fuel and oxidizer, with another 5-15 percent or so its structure; the remaining 5 percent or less is the payload. So rockets whose engines have a higher specific impulse can carry more payload into orbit and move faster in Space than those with lower ones, simply because they are more efficient and use less fuel and oxidizer.

Liquid hydrogen (LH2) and liquid oxygen (LOX) are the best fuel and oxidizer for chemically propelled rockets because of their high heat content and low molecular weight of 18 (2 and 16 respectively). They have a specific impulse of ~450-seconds. They became operational in the 1960s, principally with the J-2 engine used on the upper stages of the Saturn V Moon rocket. Its first stage used the F-1 kerosene/oxygen engine with ~330-seconds of specific impulse. The harbinger of the Space Age, the German V-2, used alcohol/oxygen engines with a specific impulse of ~230-seconds.

Solid rocket fuels appeared in the 1960s in the Polaris and Minuteman missiles and have a specific impulse of ~330-seconds. The space shuttle uses solid rocket boosters and LH2/LOX engines. NASA is developing the Ares I and Ares V vehicles, the former to boost crews into orbit, the latter cargo payloads. Both use solid rocket boosters in the first stage and an improved J-2 engine, the J-2X, as the second stage, with specific impulses of ~330 and ~450-seconds. The Ares V, the size of the Saturn V, will boost 414,000-pounds to LEO and will fly in 2018. Both are expendable or "one and done." They will not be reused.

I will further illustrate specific impulse by using the rocket scientist's term "mass ratio," the ratio of a fully fueled rocket to an un-fueled one. The space shuttle has a mass ratio of about 16 and other rockets 8-20. Here most of the rocket is fuel and oxidizer. Getting mass ratios ever closer to 1 means higher and higher specific impulses, as more and more of the rocket is payload and structure, with fuel and oxidizer weighing less and less. A related term is "payload fraction;" it is calculated by dividing the weight of the payload by the weight of a fully fueled rocket. For example, the space shuttle weights 4.5-million pounds at takeoff and carries 63,500-pounds to LEO, giving it a payload fraction of 1.4 percent. Decreasing mass ratios and increasing payload fractions means an ever more efficient and economic rocket. That is what ever-increasing specific impulse means. With chemical rockets, I have already noted a permanent barrier exists of ~450-seconds; what that then

means is forever high mass ratios as well as forever low payload fractions. For example, the developing Ares I and V will have payload fractions around 2 percent. I close this discussion, hoping I have not offended the expert, but hoping even more the general reader now understands that "While it's rocket science, it's not really rocket science, but it's really just plain old common sense. It's all about heat and weight."

Now I will discuss how rocketeers around the world have worked to get around this ceiling, to little avail. Some have proposed exotic propellants that promise specific impulses slightly higher than 450, but these are more quixotic than exotic as they are notoriously difficult to handle, or toxic, or both. Others have considered various means such as carrier planes or even balloons to lift a rocket to around 100,000-feet where it would then separate from this vehicle and go into Space. These are valid, as at 100,000-feet the vacuum of Space can be said to begin, and this eliminates much of the drag from the atmosphere and allows more efficient nozzles to be used (they perform better in a vacuum than in the atmosphere). Some launch from ships to eliminate the high costs of fixed installations such as Cape Kennedy, and launch from the equator to get an added boost. Like the previous one, this, too, is valid; both offer modest improvements but both suffer from the fact their rockets are "one and done." No reusability. Others say it's the government's fault, inefficiently giving money to the big boys of Space.

Instead, let the dynamic little boys with nifty-swifty new ideas and flashy-dashy business models solve the problem. Give them the money. Well no, the problem isn't the government or the big boys. It's heat and weight, and if they wish to spend their money on dreamy schemes that is their privilege. Our government must not continue to spend our tax money foolishly.

However, most of the effort over the decades to make gains in a rocket's performance came from using better materials (e.g., lighter weight tanks), more efficient components (e.g., better pumps and nozzles), denser fuels (saving tank weight), more economic manufacturing methods, more effective payloads (e.g., ultra-miniature and ultra-high technology ones) or more streamlined business models. Most of those gains have been realized, and those that follow will just be nibbling around the edges. This means rockets will forever have specific impulses stuck at 450-seconds, and be able to carry less than 5 percent of their take-off weight to LEO. That is quite expensive. To go beyond LEO it means rockets must carry large amounts of propellants, whose weight cannot be miniaturized; that means either carrying more fuel and oxidizer to LEO or making the payloads as small and light as possible – e.g., ultra-miniature and high tech. Both are expensive. This also means travel beyond LEO will be slow, around 25,000-miles per hour. That adds more expense: For manned missions it means more life support and radiation protection and that then requires more propellants; for unmanned ones it means waiting months or years for the payload to arrive.

This sheer expense has an obvious policy consequence: Any space program based on chemical propulsion will be forever elitist and inequitable. Only a few benefit though the rhetoric says something exactly the opposite. It says we came in peace for all mankind to land on the Moon, yet a citizen must pay $30 million to go to LEO. That is elitist. The rhetoric cites the good the space program has done: Scientific knowledge, weather and communications satellites, GPS and so forth. That is true but inequitable. Taxpayers paid for most of this, but now they must pay again for the service or be content with the knowledge flowing first to the scientists and then trickling down to them. Scientists get the awards, but John Q. Public gets the bill. This sheer expense has yet another consequence: The space program will always be one of exploration, and all thoughts of really productive space stations, lunar outposts and settlements and manned Mars missions will be flights only of whimsy and fancy. They are simply too expensive, particularly when many countries are struggling to pay for their social programs. So man remains tethered to Earth no matter how many space policy reviews are conducted, presidential statements are made, X-prizes are awarded, or rah-rah startups with razzle-dazzle business plans and nifty-swifty ideas burst upon the scene. They cannot trump heat and weight; they cannot invalidate the rocket equations. I will sum up this proposition with a colloquial analogy: When faced with the obvious need for a bigger hammer, the answer was to gold plate the nail.

## Proposition 4:
## The Nuclear Hammer

My fourth proposition is simple: The nuclear rocket is the bigger hammer. Or I can say it with "buzz-words:" The Space Age has stagnated; to prosper again it must enter the Nuclear Age. Or I can state it formally: The nuclear rocket engine is not constrained by heat and weight, like the chemical rocket engine, but by the state of knowledge at a given time. As this improves, the space program can be transformed into an inexpensive, egalitarian and equitable one. In other words, the nuclear rocket democratizes the space program and extends man's dominion from twenty-miles up to the end of the solar system – that pithy phrase democratization and dominion

– a consequence, as I said earlier, more dramatic than any prior epoch-changing shift in human history.

Before analyzing this, I must state that the term nuclear rocket does not mean it is a nuclear weapon or bomb or that it can turn into one. The technologies are fundamentally different.[ii] Rather, it signifies using the energy within the atom for propulsion, and so it is shorthand for nuclear rocket engine or nuclear powered or propelled rocket. I will just use the term nuclear rocket. Also, the term is generic, as many types exist. To cover this, I introduced the concept of the nuclear continuum in my first book.[iii] It's worth repeating, but I must caution that the word "continuum" may be somewhat misleading. It does not mean a seamless transition from one type to another is possible, but a progression in which permanent gaps may exist, as some concepts might not be feasible.

I will use this table to clarify my

| The Nuclear Continuum [1] | | |
|---|---|---|
| **Type** | **Specific Impulse** | **Comment** |
| 1. Solid core | 825-1200, or higher? | Demonstrated technology. |
| 2. Liquid core | To 3000? | Research inconclusive as to its practicality. |
| 3. Gas core | To 8000? | Research inconclusive as to its practicality. |
| 4. Fission fragment | Up to millions? | Many concepts - Orion type but all appear impractical. |
| 5. Fusion | Up to millions? | Many concepts, but fusion has yet to be demonstrated |
| 6. Antimatter | Up to millions? | Extreme radiation may make all concepts impossible. |

[1] I exclude nuclear electric propulsion concepts from the continuum because while they have extraordinarily high specific impulses, into the millions, they thrust electrons out a nozzle, and they have little mass. This means the rocket will accelerate quite slowly and be limited to carrying very light payloads.

fourth proposition, but I'll make several general observations first. The most obvious is that nuclear begins at 825-seconds of specific impulse, about twice that of chemical rockets. While good, once 1000-seconds is reached, interesting things happen to the space program; these I will discuss subsequently. Next, many nuclear concepts use hydrogen as the coolant/propellant. It is not burned with an oxidizer, just heated, and has a molecular weight of 2, so now ends one of the problems creating the impenetrable ceiling. Weight is no longer a factor. Hydrogen is the lightest element. I will say more about it subsequently. Finally, the six categories of the nuclear continuum have a collective difference with chemical rockets that burn molecules of fuel with molecules of oxidizer. Nuclear uses the energy within the atom, binding it together, and this can yield temperatures to millions of degrees, the temperatures of stars. From this, it becomes quite apparent unlimited amounts of heat are available for propulsion.

This sounds wonderful, but how this heat is contained is one limiting factor, and that depends on our knowledge at a given time. For a first-generation solid core (where the uranium fuel is mixed with another material such as graphite) that is 2000° C or 825-seconds of specific impulse. This is demonstrated. Given the advances in materials since Rover/NERVA's cancellation, advanced solid cores could, almost certainly, operate higher than 3000° C or higher than 1000-seconds.

There appears to be room for growth to 1200-seconds, though that would be problematic, but going beyond seems impossible because all known materials lose their integrity around 4000° C. To make this a brick wall, however, is unwise, as startling discoveries happen in research and development.

At one time, very early in Rover's history, some thought the solid core might reach 1600-seconds. Nevertheless, for right now, 1200-seconds looks like the solid core's uppermost limit, and even getting there would not be easy.

To get around the solid core's materials barrier and to move along the continuum, nuclear rocket designers invent different schemes to contain the uranium at increasingly higher temperatures. This pushes specific impulse ever higher. With liquid cores, the uranium is allowed to melt, and the liquid is contained in a pot-like structure similar to molten iron in a crucible. With gas cores, the uranium increases in temperature and pressure to become a plasma gas (like water turned into steam) that is contained in various ways, in some cases a bottle-like structure similar to a fluorescent light bulb. Some fusion concepts propose the use of a magnetic field to contain the heat. Some fission fragment concepts seek to use the debris from nuclear explosions, eliminating the need for a propellant like hydrogen altogether. This is Stan Ulam's Orion nuclear-pulse rocket, and it has occupied the attention of space enthusiasts for decades. Then there are sons of Orion, concepts such as using laser fusion to detonate a pellet in a chamber, the debris then exiting out the nozzle at extremely high specific impulses. All these remain theoretical for now. In sum, the ever higher specific impulses in the table mean the rocket engine is operating at ever higher temperatures, ultimately to million of degrees.

Now ever-higher temperatures yielding ever-higher specific impulses mean rockets with mass ratios getting closer and closer to one and increasingly larger payload fractions. In simpler words, it means rocket engines with greater speed *and* thrusting power. The greater speed has significance well beyond faster trips to Mars or Pluto, which engineers would view in terms of lessened requirements for life support for manned missions that then allows more for the mission itself – to stay longer and do more extensive work on the planet. For the rest of us, however, it allows us to establish boundaries on the solar system, one of the first requirements of dominion. This psychological transformation is quite important. We mentally will cease to think of it as a vast and dangerous abyss because it will increasingly shrink in our minds, from a vast Pacific Ocean as with chemical rockets to a *Mare Nostrum* with the solid core, our sea as the ancient Romans called the Mediterranean ocean, then a great lake, then a large pond, and ultimately just a nuisance puddle. In other words, progress along the nuclear continuum will cause a mental or psychological shift in which the solar system's time

and distance dimensions are increasingly less forbidding while, at the same time, our sense of personal ownership of and dominion over the bodies in the solar system increase. For example, for some, this has already happened. Some gas core proponents note 8000-seconds means Mars in a month instead of a year or more with chemical propulsion. Others hold any nuclear rocket reaching 1500-seconds opens up the inner planets for colonization while others say 2000-seconds. It is a legitimate subject for debate. Our successors in a future century may include the captain of a fusion-powered spaceship who views the now demoted planet Pluto as only a 'planetary' warning to slow his ship down as he returns from a venture beyond, and not as the outermost 'planet' of our solar system, which is our view of it now. Moreover, as this progress continues, it allows much fuller two-dimensional thinking to emerge, using Space to solve problems on Earth such as war – Szilard's main theme – instead of science, commerce, and national security as with chemically propelled rockets. In sum, increasing specific impulse means increasing the speed at which the rocket can travel, ultimately to a large but still unknown fraction of the speed of light.

The other aspect is increased thrusting power – the engine can propel ever-larger payloads with ever-smaller amounts of fuel. This says that engineers can now design rockets the way automotive or aircraft engineers do – trading off speed for power. We have automobiles for fast personal transport and slower, more powerful trucks for heavier loads. It is the same with the aircraft industry. Perhaps in the very near future, rocket designers will follow their automotive and aircraft counterparts and develop slow, powerful nuclear freighters as well as large, fast, manned transports. However, nearly all mission scenarios, up to now, feature a nuclear rocket starting in LEO and going beyond. As I noted in the opening sentence of the last chapter, this has been a conscious decision, a taboo of NASA since 1960. Out of fear of radiation, a possible "catastrophic" accident, and a public backlash over a "radioactive reactor flying overhead," NASA leaders have consciously overlooked the most significant part of this increased thrusting power: Using nuclear rockets to go to LEO. Given the times (pervasive large-scale worldwide protests against atmospheric nuclear weapons testing by the United States and the Soviet Union) this NASA decision seems reasonable, but it was pure emotion running away from technical reality, and it has had a devastating effect on the space program.

This taboo has doomed two generations of scientists and engineers to make-work efforts, when they know more than anyone the limits imposed by heat and weight. They know full well the rocket equations. It has cost taxpayers billions in pursuing marginal research and development programs to wring a few extra seconds of specific impulse from chemical rockets. With the Ares I and V, we are spending billions to go back to the past, to do what we did in the 1960s. It has thwarted the true emergence of two-dimensional Earth-Space thinking, something start-

## Origins of the Ban

In 1960, NASA and AEC had prolonged arguments over the nature and scope of the nuclear rocket program. One clash was whether to use it to reach LEO or to go from LEO. Their own words summarize that dispute. T. Keith Glennan (Administrator of NASA): "The (AEC) commissioners want to use it (a nuclear rocket) as a first stage rocket vehicle. Just where one would launch a beast with its ever-present possibility of a catastrophic explosion resulting the spreading of radioactive materials over the landscape is not clear." [T. Keith Glennan, The Birth of NASA: The Diary of T. Keith Glennan (Washington: NASA History Series, 1993), p. 73. Compare John W. Simpson (a Westinghouse vice president): "On the basis of directions of SNPO (the NASA-AEC office developing nuclear rockets), we (Westinghouse) never contemplated a launch (of a nuclear rocket) from the ground mode, all operations intended to commence after a chemical boost put the engine/vehicle into orbit. Neither we nor the AEC could see any insurmountable problems with a launch." [John W. Simpson, Nuclear Power from Underseas to Outer Space (LaGrange Park, Illinois: American Nuclear Society, 1994), p. 425. The KIWI-TNT test in 1965 showed Glennan's concerns were more emotional than technical.

ed with Sputnik, but limited ever since. It has witnessed the precipitous decline of the space industry, as firms merge to stay alive, all competing for the shrinking federal space dollar or crumbs from the private sector, and every merger means a loss of jobs and a shrinking tax base. Moreover, firms who provided goods and services to the space companies suffer a loss of jobs, which further shrinks the tax base. All these jobs haven't moved overseas, they're just gone. It has seen "innovators" arise, convinced their fresh approaches and new business plans are the low-cost ticket to LEO. No! Nifty-swifty does not overcome heat and weight, it does not replace the rocket equations, and persevering in their schemes will only get people killed, as already has started to happen. It has forced attention to illusory and magical ideas, a new alchemy, to reach LEO economically that simply won't work.

The taboo has stifled innovation

## Other Means of Reaching LEO

As noted, chemical rockets cannot take payloads to LEO economically because of heat and weight; recognizing this, many have conceived of other ways to do so, many of which do not involve rocket propulsion. Here some assume the way to escape Earth's gravity well is to ignore the rocket equations by keeping the power source on Earth and using it to propel a payload to LEO. This eliminates 80-90 percent of the propellant weight of chemical rockets. Such concepts include space cannons of different sorts – this going back to the nineteenth-century French writer Jules Verne – and space elevators. Other concepts include different magnetic or laser propulsion devices, anti-gravity schemes and even fancy slingshots and, of course, many ideas from Hollywood – transporters, warp drives and so forth. All have common features: None has ever reached the stage where its feasibility has been demonstrated and none ever will, at least not for a long, long time, as they all involve monstrously difficult engineering, magical science, or Hollywood science. In other words, they have not shown the ability to overcome Earth's gravity well and carry humans and payloads to LEO. For example, an astronaut shot to LEO by a space cannon would be squashed to a pulp by the tremendous g-forces involved, and the other concepts are tremendously expensive or require a total revolution in physics. While that may happen, while someone may come along and prove Einstein and the other giants of twentieth-century physics wrong, while someone may make the rocket equations obsolete, that may take centuries to occur. Unless and until that happens, the rocket and its equations remain the only realistic way to reach LEO and go beyond.

and creativity with all the new things that can come from a low cost means of reaching LEO. In other words, the high cost of reaching LEO makes Space predominantly reliant on public funding, not private, causing manned space activities to be justified on the basis of prestige, and limiting unmanned ones to those with proven profitability or exceptional scientific merit. It has caused public support for the space program to wane, as the average person is neither personally involved nor able to profit from it. Worst of all, it has prevented political leaders from using the space program to achieve a much wider array of domestic and foreign policy objectives. The reason why is clear: Chemical propulsion only gives, in reality, about 2 percent payload fractions and that is quite expensive, as the ammunition philosophy rules. One and done. However, moving along the nuclear continuum means an increasing larger payload fraction, from perhaps 8 percent of gross takeoff weight for a first-generation nuclear rocket to more than 90 percent for fusion. Unless this taboo is broken, the space program will rely, *ad infinitum*, on chemical propulsion to reach LEO and take payload fractions there of less than 5 percent, *ad infinitum*. The cork will remain in the bottle and the space program will remain grossly elitist, expensive and inequitable. This is heat and weight talking – the rocket equations - on the consequences of being ignored.

My technical discussion of nuclear

---

### The Space Program, Nuclear Rockets and the Pony Express

The Pony Express lasted from April 1860 to November 1861, and bears a remarkable resemblance to a space program conducted with chemical propulsion. Each pony weighed less than 900-pounds, each rider under 110, leaving 55-pounds for the saddle and bridle, rider's clothing and a 20-pound mail sack. The total: 1065-pounds, or to use space-age terminology, the Pony Express had a payload fraction of 1.8 percent, quite close to the space shuttle's 1.4 percent and the ~2 percent for the Ares I and V. The new hammers of the time – the railroad and telegraph - doomed this enterprise. If the 1960 taboo continues, prohibiting a nuclear rocket's use to reach LEO but allowing it to fly from LEO to deep space, a ludicrous situation develops, akin to requiring the Pony Express to take mail from St. Louis to San Francisco, where a Boeing 747 would pick it up and fly it to Tokyo or Beijing.

---

rockets has been marvelously glowing, but now I must add some hardheaded realism. If our knowledge of how to contain the heat is one limiting factor, the other critical one is how to start and control the nuclear process. In the case of solid, liquid, and gas cores, the heat derives from the fissioning of the uranium atom, which then heats a propellant, which then leaves the nozzle at a high velocity to produce the forward thrust. The fission process is well known and understood. Some fission fragment concepts use the debris (neutrons, gamma rays, and fission fragments) from a nuclear detonation to propel the rocket (this debris travels somewhat below the speed of light). This is most identified as the Orion nuclear-pulse rocket, and during its existence the researchers reduced the size of the nuclear explosive dramatically, in essence making it a shaped charge. However, Orion itself is not feasible because of insuperable materials problems. Moreover, the technology to make nuclear-shaped charges remains highly classified and unavailable publicly. Other sons of Orion, however, would use a fusion reaction, such as lasers striking a pellet inside a chamber, exploding one after another, with the debris exiting out the nozzle. Other laser fusion concepts are similar to the solid cores that heat hydrogen. Laser fusion, however, is a not-yet technology. With fusion, the nuclei of lightweight elements are combined to form heavier and more tightly bound nuclei and release vast amounts of heat. The use of fission to trigger a thermonuclear reaction is well understood but not well known, because of classification, and is popularly called an H-bomb. However, that is uncontrolled. For the last fifty years, scientists in many countries have worked to achieve a controlled fusion reaction. The last is antimatter, the annihilation of matter to produce prodigious amounts of energy; it is a developing field of scientific inquiry and any thoughts of using it for rocket propulsion are more fiction than science at the present time. Moreover, its extreme radiation may prevent it from ever being used.

From this review, I draw several conclusions. The solid core is demonstrated, but experts differ on the feasibility of liquid and gas cores, a view that also applies to some fission fragment concepts. These have suffered from an utter lack of research and development, so in principle, their feasibility might be demonstrated with increased attention.[iv] For all fusion concepts, however, the knowledge to control the heat as well as to initiate and control the reaction is lacking, but breakthroughs can occur at any time. Considerable sums of money are being spent on fusion worldwide, and many good scientists are diligently making good progress. Once a breakthrough occurs, however, time and effort will be required to adapt it for propulsion, but it is impossible to speculate with any confidence now on the time and cost involved. Finally, antimatter is too theoretical to be of any use for propulsion now. In sum, a continuum of ever-increasing capability exists for nuclear rockets, with ever-increasing speed, thrusting power, and specific impulse that can ever-increasingly change the space program. But currently only a nuclear rocket at the beginning of the continuum is demonstrated conclusively: The solid core.

## Proposition 5:
## The Nuclear Hammer's Infrastructure

My fifth proposition is the nuclear rocket will require its own infrastructure to achieve its potential, but some developed for the Nuclear and Space Ages may be of use. I will carry out this argument in the succeeding chapters, but here let me review the infrastructures that have developed for the Nuclear and Space Ages and for Project Rover/NERVA.

In general, the origins of the Nuclear and Space Ages can be traced to around 1900 with the work of atomic scientists trying to unravel the mystery of the atom's structure and with the experiments of the rocket pioneers. Both groups worked independently, and both gathered momentum during the 1930s. Both came to fruition during WWII with the V-2 rocket and the detonation of three atomic bombs in the summer of 1945. Here, though, that division solidified. On the one hand, Space Age infrastructure continued to develop as the United States and its industries poured resources into rocket development for civil and military uses (to deliver nuclear weapons), and as scientists justified sending satellites into orbit or writers glamorized what could be done in Space. This decade-plus of activity prepared the public, so when Sputnik occurred in 1957 the U.S. reaction was not "So what!" or "Who cares?" but "Why are we lagging?" and "What can we do?" What was done was the creation of NASA and the further "concretization" of the space program with more research and development facilities, private companies, mission operations, and launch complexes, all of which yielded the lunar landings, the space shuttle and International Space Station, as well as scientific and commercial satellites. Congress likewise reorganized itself with different House and Senate committees exercising oversight on this infrastructure.

Overseas, many countries followed this pattern, creating space agencies and aerospace industries, and adding research and development and launch and mission operations complexes. Except for the Soviet Union, governments also mandated this activity be public, minimizing classification and secrecy, and cooperated in the space programs of others, offsetting the high cost of space operations. Thus, a worldwide infrastructure exists to support and justify the space program, so when an accident happens, such as debris plummeting to Earth or the unfortunate loss of life, the public does not say "Stop the space program!" but "How can we do it better and safer?" All of this, of course, centers on chemical propulsion.

On the other hand, the Nuclear Age followed a different path. It started, of course, for military purposes, and that continued with the creation in 1946 of the Atomic Energy Commission. When the Soviet Union detonated its first atomic bomb in 1949, a nuclear arms race began, which accelerated after the decision to build the H-bomb in 1950. The infrastructure that developed in the United States was primarily governmental, with a heavy dose of guards, gates, and guns, and security and classification rules. What the public saw of this complex were many atmospheric nuclear weapons tests, and what they heard were government assurances of nothing to fear from radioactive fallout. Not all agreed, and the press carried story after story about fishermen and South-Sea islanders suffering and dying from fallout, children drinking radioactive milk, or Laplanders eating radioactive caribou. Government apologies had little effect, and protest movements drowned out any scientific rebuttals.

In 1954, the Atomic Energy Act was amended to permit private sector involvement, including the right to own nuclear materials under a government license. This allowed a civil infrastructure to develop in the United States (and permitted the United States to catch up with the rest of the world), with a selected de-emphasis on classification and security, but still with a heavy dose of guards, gates, and guns. Much of this activity centered on the light water reactor. A hundred of these produce one-fifth of United States electrical energy requirements. The revision to the Act also permitted the United States to cooperate with other nations in peaceful nuclear activities under carefully delineated conditions; this allowed the light water reactor to dominate overseas.

Finally, the change led the United States leadership to create the International Atomic Energy Agency (IAEA) in 1957; it would exercise a carefully delineated right, but not authority, over a member nation's sovereignty through its safeguards system. That would allow it to inspect, and assure all other states, that the subject country was not diverting its program to malevolent purposes. Nearly every country joined the IAEA, and most created an infrastructure, some modest but others robust with a full array of facilities to support a vigorous program, with an emphasis on electrical power production.

This started the creation of an international infrastructure to govern the atom, and in 1963, the United States and the Soviet Union signed a treaty banning nuclear testing in the atmosphere, underwater, and outer space, but allowing it underground. In the following years, the two signed other arms control treaties though now it can be said that the Non-Proliferation Treaty (NPT) is the kingpin of them all. Signed in 1968, the NPT has nearly every country in the world as a party and features a two-part bargain. The non-nuclear weapons states forswear their right to develop nuclear weapons in exchange for access to peaceful nuclear technology, while the nuclear weapon states are obliged to reduce and ultimately eliminate their nuclear arsenals in the context of general and complete disarmament.

In 1955, the United States' nuclear rocket program appeared, and it was highly classified, though by 1973 that classification had been downgraded significantly. Rover/NERVA developed an infrastructure surrounded by guards, gates, and guns and restricted public access: Los Alamos, the Jackass Flats area at the Nevada Test Site, and Y-12 in Oak Ridge, Tennessee. The industrial contractors such as Rocketdyne and Aerojet developed non-nuclear parts such as pumps, nozzles and engine controls, so security there was not as stringent. The exception here is Westinghouse, which had an AEC license to develop reactors and test fuels at its own facilities around Pittsburgh.

When the program ended in 1973,

---

### The Alleged NUMEC Diversion

Westinghouse operated two fabrication facilities to provide fuel for its nuclear rocket reactors. At one, the NUMEC plant in Apollo, Pennsylvania, there was an alleged diversion to Israel of significant quantities of high-enriched uranium (HEU). The issue centers on material unaccounted for (MUF). In all manufacturing operations, some material is lost in making the finished product, and making fuel rods for nuclear rocket reactors is no exception. Some hold the HEU was just lost here, a MUF, while others allege a purposeful diversion occurred. The government conducted highly classified reviews that concluded no diversion occurred; however, not one, to my knowledge, has been released, so rumors of a diversion persist.

---

the technical infrastructure at Aerojet, Westinghouse, and Rocketdyne vanished while the government infrastructure disappeared via cannibalization, decay, or demolition, particularly at Jackass Flats. Moreover, many original documents seem to have been lost or destroyed, though I must say determined efforts have recovered much of the core documentation. Today's infrastructure consists of pockets of experts who write reports on different nuclear concepts with little research and development to guide those efforts. I conclude the nuclear rocket's infrastructure is virtually nonexistent. Even more significant, Congress abolished the JCAE in 1975 and dispersed the functions of this political infrastructure among many committees of the House and Senate, leaving the appearance of everyone in Congress overseeing the nation's civil and military nuclear programs. There is no "Mr. Atomic Energy," the title once given to Congressman Chet Holifield of California, though others deserve that designation as well. Even worse than that, there is no acknowledged forum such as the JCAE for debate and discussion, for the initiation of legislation, and for oversight under the *"fully and currently"* provision (this clause obligated the AEC to inform the JCAE "fully and currently" of its activities, almost on a day to day basis). Perhaps worst of all, there is no dedicated committee staff with its detailed knowledge and long corporate memory of atomic energy affairs.

The demise of this joint committee has contributed to the coarsening of public dialogue over atomic energy. The 1980s saw bitter protests over the proposed development of a neutron bomb that kills living things but leaves structures intact. It continued with nuclear winter scenarios, a full-scale exchange between the United States and the Soviet Union creating a massive cloud that some said would blot out the Sun, and the nuclear freeze movement, to cease production of all materials for nuclear weapons. The Chernobyl and Three Mile Island reactor accidents contributed to the bleak state of affairs. With no single congressional forum, the executive branch had to respond, and bureaucrats do what bureaucrats do best: Explain and then hunker down and wait for it to blow over. But it doesn't blow over. It only subsides, as the Pentagon realized when it tried to develop Project Timberwind, a nuclear rocket for the Strategic Defense Initiative. The press learned of the secret program, and controversy quickly followed that helped end the effort. NASA realized it when it proposed to develop a nuclear rocket for the Space Exploration Initiative of President George H.W. Bush. What followed was a proposed treaty in the United Nations (UN) to limit a nuclear rocket's use in Space, to where it was "safe" and could not "harm" the Earth. Wisely, the United States withdrew its support from that effort. NASA learned of it again when it proposed a plutonium-238 electric generator for the Cassini mission to Saturn in the 1990s. Protests followed by those who felt it posed a grave danger to Earth.[v] In sum, had there been an acknowledged forum, such as the JCAE, the debate would have still been heated, which is acceptable, but not vituperative and confined to a 15-second sound bite on the evening news. The public would have thereby had a fairer opportunity to judge the better argument.

From this I conclude a worldwide infrastructure exists for the space program based on chemical propulsion, but the nuclear infrastructure centers on power plants and weapons. None exists for the nuclear rocket. I hold it must be created to fulfill this technology's potential. In other words, if trains need tracks, automobiles roads, airplanes airports, and telephones poles, then the nuclear rocket needs its own infrastructure, and creating that must be the primary aim of a reconstituted program, not some type of mission. Those will come – after an infrastructure exists to use nuclear rockets effectively - and allow it to democratize the space program and extend its dominion to the end of the solar system.

## Proposition 6:
## The Law of Supply and Demand

My sixth proposition is that the law of supply and demand applies in Space just as it does on Earth. In other words, as the cost of reaching LEO and going beyond decreases, demand for this service will increase. I believe few will dispute this: It's Economics 101. This proposition, however, has two corollaries. One says as costs decrease, the private sector will increasingly fund the activities it sees as desirable or profitable; how this is structured is debatable. The other says as this occurs, government space programs must transition to a different model, like the one for the airlines, automobiles, and railroads, where the government funds only certain activities such as safety, then regulates, and promotes competition and fairness. Breaking the taboo will make both corollaries a reality, and I will offer my thoughts to begin the debate over these new structures.

## Summary of Propositions

I first proposed technology has consequences, but stated that the most epoch-changing of all – from Stone to Bronze to Iron to Machine/Industrial to Nuclear to Space Ages – the most rev- olutionary of all will be the nuclear rocket.

Second, I noted technologies require an infrastructure in which they

---

## The Most Epoch-Changing of All

Perhaps this thought is what Stan Ulam had in mind when he said the nuclear rocket was more revolutionary than the H-bomb he co-developed, but if this is so, I must qualify it. The Atomic and Space Ages began fero- ciously and unexpectedly for most of the peo- ples of the world, frightening and startling them and their governments. However, a nuclear rocket's changes will be more grad- ual, particularly as progress is made along the continuum. Yet when historians look back, they will point to the restarting of the program as the beginning of a new epoch. To illustrate this further, an example from the shipping industry is insightful. Weighing 22,500- tons, the Great Eastern was a monstrously big, iron-hulled, steam-powered ship nearly 700-feet long that appeared in England when graceful clipper ships ruled the seas. Launched in 1858 (well before the famed Monitor and Merrimac of the United States Civil War), scrapped in 1889, and never prof- itable, the Great Eastern still represented the future, and by World War I, iron-hulled ships plied the oceans. The great age of sail was over. The transition took about fifty years. I expect something similar to occur with the nuclear rocket, though perhaps over not so long a time span - maybe only twenty or thir- ty years.

---

develop and which governs them. Third, I concluded the Space Age has stagnat- ed because of heat and weight, leaving it permanently expensive, elitist, and inequitable. It needs not more gold-plat- ed nails, but a bigger hammer. Fourth, I said the nuclear rocket is the bigger hammer, but only the solid core is demonstrated. Fifth, I held it requires its own infrastructure, a discussion that may seem remote from the nuclear rock- et, but is not. If they allow man to extend his dominion into the solar sys- tem, an infrastructure must exist to sup- port and manage it and ensure that it is egalitarian and equitable. Here I cate- gorically differ from those who hold the mission must be defined first, then a nuclear rocket developed for it. This is simple "mission-itis" thinking, some- thing appropriate for single-shot chemi- cal rockets. Here it is enough to recog- nize that breaking the taboo will change the economics of moving into and around Space, my sixth proposition, and that alone, without any mission, requires a complete rethinking of the space pro- gram. The next seven chapters will start that, and hopefully from the discussion and dialogue they create, a detailed jus- tification for the program will emerge. The proofs of its success will be vision- ary political leaders who emerge to sup- port the nuclear rocket as well as citi- zens who will now demand to partici- pate and benefit, no longer chanting "Stop threatening our planet." Thus, the task is more complicated than simplistic mission thinking because it centers on infrastructure, and my analysis of that begins in the next chapter with my assumptions about the solid core, the first of the bigger hammers.

[i] Richard G. Hewlett and Oscar E. Anderson, <u>This New World, 1939-1946: A History of the U.S. Atomic Energy Commission</u> (Pennsylvania: Pennsylvania State University Press, 1962).

[ii] The exception is the Orion nuclear pulse rocket; in theory, it would detonate small nuclear explosives at the base of the rocket to propel it forward. In practice, however, Orion cannot work because of insuperable materials problems.

[iii] <u>To the End of the Solar System</u>, pp. 237-238, 253.

[iv] NASA funded gas core research at a million per year, but that ceased in 1978. That was too little to make a definitive statement on a gas core's feasibility.

[v] Karl Grossman, <u>The Wrong Stuff: The Space Program's Nuclear Threat to Our Planet,</u> (Monroe, Maine: Common Courage Press, 1997).

# Chapter 3
## Three Assumptions

Since Rover/NERVA ended in 1973, designers have studied many solid core concepts and touted their advantages for this or that mission. As paper studies, conducted without the benefit of research and development to guide the effort, they are suspect. In contrast, starting in 1955, Los Alamos went through a chicken and egg process, with experiments to guide designers, and with developing designs to challenge experimenters. By 1960, that produced a broad, detailed literature base that led to the B-4 core. However, in the 1980s the Pentagon disregarded it and established Project Timberwind, a nuclear rocket for the Strategic Defense Initiative. It was based on a core concept Los Alamos studied in the late 1950s, called Dumbo, and discarded as unworkable on the basis of fundamental laws of physics and hydrodynamics. Only now, $400 million was spent to prove it. Composed of ball bearing fuel elements packed together, the core could not be cooled adequately, allowing hot spots to develop. Avoiding them meant running at lower temperatures, reducing specific impulse; ignoring them risked bearings increasing in temperature until they failed, creating a domino effect until the core came out the nozzle.

### Assumption 1:
### Picking the B-4 Core

This assumption is the most impor-tant: I presume a reconstituted program will not reinvent the wheel and will select the B-4 core. I have five reasons for this, all-important for my argument, but will discuss just the first two here. The other three are in Appendix A, which summarizes B-4 development.

First, twelve years of intense effort (1960-1972) went into the B-4 core, and it only had to complete the stage development phase to be declared operational. No insuperable problems exist here. It will work. Period. However, starting over on new or so-called improved designs risks core failures, and that leads to negative publicity and criticism. The conflict with President Kennedy's science advisers after the early KIWI-B1 and KIWI-B4 core failures as well as the ill-advised Project Timberwind offer an instructive lesson: Learn from history.

My second reason is that the program must focus on building the infrastructures in government, the space industry and the public, creating new ones or adapting those already in existence to the revolutionary consequences of nuclear rockets. That will occur most easily by moving to the flight test and operational phases with a first-generation system as soon as possible, and building from there. In contrast, concentrating on new core designs for this or that technological advantage shifts the focus to proving its soundness, and every problem will create management and technical reviews. They bring hesitancy and doubt, resulting loss of time, and this risks loss of momentum.

I offer historical and theoretical examples to justify these points. In the late 1940s, the navy concentrated on developing a robust yet simple nuclear reactor and integrated it into a hull, and this submarine had torpedoes as its military armament. Launched in 1954, the Nautilus, however, did not have torpedo attack as its primary mission. Rather, it was a prototype, a learning tool, from which infrastructures developed within the navy, then within the larger defense and foreign policy sectors. Designers quickly came up with better hull shapes to take advantage of the reactor's capability and of improvements in solid fuel missiles and warhead design. A fleet of Poseidon submarines emerged in the 1960s carrying Polaris missiles, something unthinkable in the 1940s, and nuclear submarine development has continued ever since to conduct missions also unthinkable in the 1940s. Operations personnel realized the need for command and control and training of crews so that infrastructure came into being. Military strategists took a vision of the 1940s to become a vital part of the nation's deterrent structure, so that infrastructure of policy and doctrine developed. Today, political leaders and defense officials rely on it in their conduct of foreign policy or in a crisis. Ultimately, this infrastructure spread to the public realm, and those cities that rejected the presence of nuclear powered ships, out of a fear of them blowing up in their harbors, probably would second-guess their decisions after viewing the stable economies of cities that welcomed this naval presence. The public now by and large accepts a nuclear navy as a vital component of the nation's defense posture. That may never have come about were it not for the uncompromising Admiral Hyman G. Rickover, the father of the nuclear navy, who kept the focus on simple and reliable reactors, and the equally uncompromising JCAE, which retired or shunted aside those hostile to Rickover. This paid great dividends during the Cold War.

### THE NAUTILUS

PHOTO CREDIT: U.S. NAVY

The world's first nuclear submarine, the Nautilus had only torpedoes for armament, a throwback to World War II thinking, but soon astounded everyone by sailing submerged around the world. The photograph shows the Nautilus entering New York harbor after sailing under the North Pole in 1958. That opened the eyes of military strategists who then conceived of many missions, such as Poseidon submarines carrying Polaris missiles. This would have been inconceivable in the late 1940s when construction of the Nautilus began.

To those who prefer a theoretical discussion, I advise reading the "Mission Merry-Go-Round" speeches of AEC commissioner James T. Ramey. He gave them to fight the "missions-first" ideology that President Kennedy's science advisers invented to kill any program with which they disagreed: Approve a mission first, then build the things necessary to do it. Unfortunately, this ideology has morphed into "mission-itis" thinking in NASA, the space industry, other corporations, and the public; the prevalent belief today is that a mission must exist before initiating development. It was wrong not only for the nuclear rocket then, and it's wrong for any research and development program now. Ramey says why: This self-professed technological liberal argued promising technology development efforts should proceed through the prototype stage where something is built for full field-testing and evaluation. That allows the nation's leaders, or corporate officers, to make more informed decisions on proceeding, yet it avoids strangling promising technologies with mission requirements, which may become apparent only after full-field testing. At the same time, it keeps the budget under control, and it begins infrastructure development to manage and operate the new technology. Here I categorically state that it is highly dangerous and irresponsible to assign missions to a new technology without knowing its handling capabilities and without having a management team fully competent in its operations. Rereading Ramey's speeches will serve as a strong antidote to "missions-first" thinking for most technology development efforts, but my argument will kill it as it pertains to nuclear rockets.[i]

## Assumption 2:
## The Four Waves of Engine Development

I have made a solid case for resuming development of the B-4 core and not reinventing the wheel. My second assumption follows from this, that a small engine based on the B-4 will be developed first, going through four-generations before developing a large engine. I further assume this development will center first on its use to reach and return from LEO – breaking the taboo – and once achieved, then a small engine is developed for uses beyond LEO. After that happens, the go-ahead to develop large engines will be given, first to reach LEO, then sometime later, to go beyond. (This should take about a decade, which may seem too optimistic, but I discuss the timeframe in Appendix B). So we have a structured pattern, like four waves flowing to a beach one after another: A small engine to LEO, then beyond LEO, then a large engine to LEO, then beyond LEO.

Gaining experience with smaller engines first, as with the Nautilus, allows infrastructures to develop and begin maturation before they must be transformed into structures for the large engine. Note this sequence is devoid of missions initially, but it is not devoid of objectives. Those will be many, including public relations ones to demonstrate safety and to attract investment capital. However, when fourth-generation systems appear, then a burgeoning increase in missions will occur, I maintain, because a more mature infrastructure exists to make effective use of them, i.e., an infrastructure now exists to promote a democratic and fraternal space program.

Offhand, I guess the casual reader would see this as so bland and innocuous that only the critics who fear nuclear reactors in Space could object. Yet it is in fact most stunning and momentous because it repudiates the prohibition against using nuclear engines to reach and return from LEO. The taboo has been broken. But it's far more than just that. It changes the basic structure of the space program, everywhere. Programs based on chemical propulsion leave the executive branch paramount, with a president, prime minister or space agency announcing space goals and the legislative branch doling out money. In the United States, Congress deals it to NASA, which distributes it to its complex and contractors. Lawmakers then oversee operations. With the four waves of engines, first small then large, the legislative branch becomes paramount. However, it's much more than that, as a fundamentally new, epoch-changing space program can be created everywhere, in which the private sector is dominant, and NASA and the space agencies of other countries are then restructured to support it. Yet it is even far more than

that, as new regimes aimed at world peace and Greening the planet can be created. I will show how in the following chapters, but here I finish my discussion by looking at the history of using nuclear rockets to reach LEO.

Earlier I quoted the statement of the AEC and Westinghouse who saw no technical problem with it, but in the late 1950s many concepts existed. This thinking started while the Air Force had the program (1955-1958) and aimed it to be a heavy lifter to propel large payloads into Earth orbit. NASA inherited the program from the Air Force in 1958, had two years of indecision on its use, then in 1960 established the ban. During this time, the Douglas Aircraft Company had its RITA (Reusable Interplanetary Transport) concepts, where a RITA-A could take 85,000-pounds to LEO or land 19,000 on the Moon, then return to Earth. A single-stage RITA-B could take 160,000-pounds to LEO or 25,000-pounds to the Moon. Douglas also developed tanker and tug RITAs for complex space operations.

Krafft Ehricke had his Helios concept, with a mammoth and utterly unrealistic 15,000 MW reactor to be a heavy lifter. However, this V-2 German who became a key figure in U.S. rocketry conceived of reusing its chemical rocket boosters. He also conceived of a two-stage piloted nuclear rocket to reach LEO. The first would use ammonia, take off horizontally from the ground, fly 30-miles up then separate and glide back to Earth for

**RITA**

The thinking behind the RITA-A concept, using a chemical rocket as a first stage, is a good model to follow. It boosts the stage to higher elevations where the nuclear rocket starts and propels the stage to LEO. What is necessary, however, is a boost to higher elevations, and that need not come from a rocket. The all-nuclear RITA-B, taking off from the ground and going to LEO, is impossible for radiation and weight reasons unless the nuclear engines have specific impulses well over 1000-seconds.

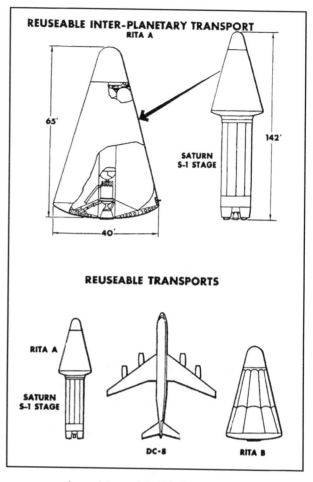

REUSEABLE INTER-PLANETARY TRANSPORT
RITA A

SATURN S-1 STAGE

REUSEABLE TRANSPORTS

RITA A

SATURN S-1 STAGE

DC-8

RITA B

reuse. The second stage would use LH2 and fly to and from LEO. The system was to weigh 350,000-pounds at takeoff with a length of 180-feet and a 50-foot wingspan. Ehricke noted radiation shielding was a major design problem, more so for the first stage and its flight crew than the second. It's an interesting concept because using ammonia in the first stage would give a specific impulse of around 600-700-seconds, double that of the first stage of the developing Ares V. He did no further work on it to my knowledge.

Such thinking ceased after the 1960 prohibition, though in the 1960s and 1970s scattered references to nuclear heavy lifters appear. (Robert W. Bussard introduced his ASPEN concepts in 1961 and 1971; I discuss them in Chapter 5). Robert Seamans, the number two man in NASA, mentioned it on his flight on Air Force One in 1962 with President Kennedy on his review of the nuclear rocket, but did not elaborate.[ii] In 1968, James Webb told Congress a 1500 MW NERVA I, as the third stage of the Saturn V, could boost 450,000-500,000-pounds into LEO, compared to the 250,000-280,000-pounds of the all-chemical Saturn V.[iii] (Compare that with the space shuttle's 65,000-pounds or with the Ares V's 414,000-pounds). However, the Saturn V disappeared a few years later, so any thought of using NERVA as its third stage also disappeared. Meanwhile, the shuttle grew in prominence and so, as always, NERVA

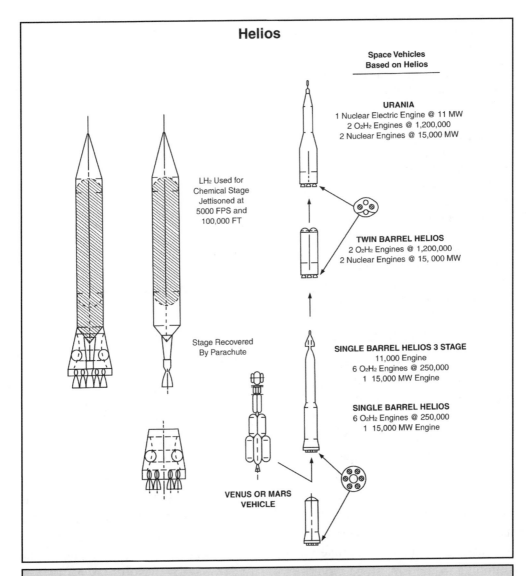

## Helios

**Space Vehicles Based on Helios**

**URANIA**
1 Nuclear Electric Engine @ 11 MW
2 $O_2H_2$ Engines @ 1,200,000
2 Nuclear Engines @ 15,000 MW

LH₂ Used for
Chemical Stage
Jettisoned at
5000 FPS and
100,000 FT

**TWIN BARREL HELIOS**
2 $O_2H_2$ Engines @ 1,200,000
2 Nuclear Engines @ 15, 000 MW

Stage Recovered
By Parachute

**SINGLE BARREL HELIOS 3 STAGE**
11,000 Engine
6 $O_2H_2$ Engines @ 250,000
1 15,000 MW Engine

**SINGLE BARREL HELIOS**
6 $O_2H_2$ Engines @ 250,000
1 15,000 MW Engine

**VENUS OR MARS VEHICLE**

## HELIOS

PHOTO CREDIT: U.S. CONGRESS

Krafft Ehricke's Helios concept featured first stage chemical rocket boosters pushing a mammoth, 15,000 MW nuclear engine that would then propel hundreds of tons to LEO as economically as possible. To keep costs low, he had the chemical rocket booster separating at 100,000-feet and parachuting back to Earth for re-use while the nuclear engine would then fire and push the payload into Space. It would stay there and be used for manned planetary missions. This gargantuan nuclear engine could not be built, but such thinking was in vogue during the late 1950s and early 1960s because all assumed a nuclear engine's fuel element could last for only 10 to 30-minutes. So to get the thrust needed, particularly for the manned mission concepts of the 1950s, a very big engine firing for a short time seemed to be the answer. Within several years, however, engine sizes dropped dramatically because the fuel could last much longer. In 1967, the NRX-A6 ran for more than an hour at full power. So the answer became building a smaller engine, and just running it longer to get the same amount of thrust overall.

was to start in LEO and carry payloads to points in the solar system, with the shuttle carrying LH2 and payload up to it. However, I neither found nor read all mission studies during this period so heavy lifters might have been studied. I doubt it though, so pervasive was the NASA prohibition.

One memorandum from December 1966 reinforces this point. A different NASA office asked the Space Nuclear Propulsion Office (SNPO, which managed nuclear rocket development) to consider using nuclear engines as heavy lifters, and the five-page response concluded a 50 percent payload gain to LEO was possible. It centered on the Saturn 1B, with a nuclear stage replacing the S-IVB second stage, and noted it could lift 54,000-pounds to LEO compared to 33,000-pounds of the all-chemical S-1B. It also considered the 260-inch solid rocket booster (which was never built) as a first stage and concluded similar payloads could be launched at reduced costs (as solid rocket engines are less expensive than liquid ones). The memo's principal shortcoming was the static thinking inherent with chemical systems. It took the 33-foot wide nuclear stage that was in vogue at the time, powered by a large NERVA II engine, also in vogue at the time, with a specific impulse of 820-seconds, 200,000-pounds of thrust, and 40,000-pounds of weight, and said: "How can we lift a payload with it to LEO?" It did not consider the problem afresh, but worried about hammer heading – the friction, drag, and instability of a

33-foot wide stage and payload sitting atop a first stage 21- or 22-feet wide. It ended dismissively: "…(I)t appears that a second-stage, non-reuse application may offer some performance advantages, but their importance is not clear. We do not plan further analysis at this time…"[iv] Ironically, this came from SNPO, which we would not expect to display such stagnant chemical rocket thinking.

## Assumption 3: Specifications of Small and Large Engines

The small-large engine split is too vague to permit any analysis: How small is small and how large is large? So I must specify their capabilities. To do so, I will use the technical specifications I expect to see in <u>fourth</u>-generation engines in my argument. Let me make some general comments first. In Rover/NERVA, the emphasis was always on large engines for manned missions, particularly Mars, yet studies of small engines dot that history. However, focus on them did not begin until the Pewee, tested in 1968 as a *fuel* test reactor. Only afterward did thoughts begin on developing it into a flight-rated system, as NERVA had run into trouble in Washington. Pewee used the B-4 core concept, though it was only 20-inches in diameter compared to the 35-inches for the KIWI-B4, Phoebus 1s, and NRXs and 55-inches for the Phoebus 2A. Yet like the others, its core remained 52-inches in length. In 1968, it was tested at a power of over 500 MW, and had a specific impulse calculated to be 845-seconds. In 1971-1972, the program's final years, the Small Engine concept evolved to fit in the cargo bay of the space shuttle (15x60-feet). It derived much from the Pewee, and was to have a power of ~400 MW (20,000-pounds of thrust), a specific impulse of 875-seconds, a weight of ~6000-pounds and a short, squat core of 37x37-inches (shorter to fit in the cargo bay). It still used the B-4 core concept. It even featured a hinged nozzle that swung back and saved 4-feet for its shuttle ride to Space, after which it was swung back and locked into place.

My sense is a small engine would look like Pewee (one tested, one built for testing and three on the drawing boards when Rover/NERVA ended) and not like the short-stubby one for the shuttle. If so, I expect when its fourth-generation version appeared it would have 800 MW of power (40,000-pounds of thrust), a specific impulse of 1000-seconds and a weight of 6000-pounds. I also expect two different models would appear, a Re-use and a Re-core. While the former has a clear pedigree from Rover/NERVA – indeed the Small Engine, like NERVA, was to have a 10-hour fuel lifetime with 60-recycles or stops and starts, making it a taxicab in LEO – the Re-core is totally new and originates solely with Robert W. Bussard, no one else ever studied it. Stunning in its simplicity and staggering in implications, the Re-core uses fuel once in a run of about 10-minutes, 15-minutes maximum, then it returns to Earth, the core

is removed, and a new core inserted into the pressure vessel, and the somewhat irradiated one recycled to recover the uranium. (All solid core nuclear rockets use highly enriched uranium, that is, uranium enriched to 93 percent U-235 content. This is a costly process. However, in a 15-minute flight, it might consume just one percent of the U-235, so recycling makes economic sense). This greatly simplifies launch operations, permitting hands-on handling, as it would be virtually non-radioactive. It will be used for going to and returning from LEO. The Re-core also changes fuel element development to emphasize retention of fission products, economics of manufacturing, and recycling. I'll discuss this more subsequently. With hesitation, I assume a fourth generation Re-use engine will have 30-hours of full power operation with 180-recycles, or stops and starts. This will be hard to achieve, but it is not out of the realm of possibility, given ten years of fuel development. The Re-use engine will operate to and from LEO, like a taxicab going to other points from LEO, and then returning. Finally, I assume a fourth-generation Re-core will boost a 20,000-pound payload into LEO; the Re-use engine will take it from there.

The large engine would be based on the 35-inch core NERVA I (though it's not the only choice possible) and I expect its rapid and swift development, leapfrogging from a first-generation system to a fourth, skipping over the second and third. The reason is simple. Strong and competent management teams will have a decade-plus experience with small engines, and the B-4 core is scalable. It can go up or down in size quickly – indeed, Los Alamos went from the 35-inch KIWI-B4 and Phoebus I to the 55-inch Phoebus 2 to the 20-inch Pewee and even to the 37x37-inch Small Engine within a few years. I expect a power of 3000 MW (150,000-pounds of thrust), a specific impulse of 1000-seconds and a weight of 15,000-pounds. Unlike the small engine, which would have Re-core and Re-use versions, this large engine would appear as a Re-core initially, with a Re-use version coming in later. I will explain why in Chapter 5. I expect this heavy lifting engine to carry 150,000-pounds to LEO.

Thus, I assume two different fourth-generation engines will appear to boost 20,000 and 150,000-pounds to LEO. I know my methodology is unusual, estimating the fourth-generation specifications of an engine that never flew to analyze how it can democratize the space program and create a new epoch in human history. I justify this starting point as follows: Little could be said about the consequences of the first railroad engines, the first automobile, the first telephone, or the first airplane. They were too new and their technology too crude and immature, and too was little known about their operating environment or infrastructure requirements. For example, how could anyone project the existence of a worldwide commercial airline industry after the first Wright brothers' flight in 1903 or even

their second-generation one of 1905 that went a staggering distance of 24-miles in 38 minutes? However, after the technology matured somewhat and a basic infrastructure existed, e.g., the railroad after 1860, the automobile and telephone after 1920, and the airplane after 1930, then an analysis of its consequences can proceed on firmer grounds. Only after those dates did they start to shape the political, economic, social, and intellectual lives of the people in whose countries they appeared. Today, a great deal is known about atomic energy and Space, and each has well-established infrastructures, and since the performance specifications of fourth-generation engines are likely, we can consider how they democratize the space program and discuss the new institutions they will require. We are not flying blind, but through a haze.

Let me summarize where we are now. In Chapter 1, I discussed many different ideas and themes derived from breaking the taboo and packaged all in the phrase "democratization and dominion." Henceforth, at the start of each succeeding chapter, I will refresh the reader's memory of each particular idea or theme to be considered. This will be in italics. In Chapter 2, I considered six propositions, beginning with the premise that technology has political, economic, social, intellectual, and religious consequences, but requires an infrastructure to develop and govern it. Then I outlined the infrastructures that developed for the railroad, automobile, telephone, and

## PLAIN WORDS ABOUT SPACE NUCLEAR ACCIDENTS

Since Sputnik, there have been eight Space nuclear accidents, three by the United States and five by Russia, in which a reactor or radioisotope thermoelectric generator (RTG) returned to Earth. (RTGs convert the heat or thermal energy of an isotope such as plutonium-238 into electricity; the fission process does not occur, as in a reactor). The first U.S. accident occurred in 1964 when an RTG failed to achieve orbit and burned up upon re-entry and dispersed its nuclear inventory to the upper atmosphere. It was purposely designed to do this in the event of a launch failure, so calling it an accident is not really accurate. Still, that caused a change in design philosophy to emphasize the containment of the nuclear material in the event of an accident. That proved successful in 1969 when a rocket launch was aborted and the two RTGs fell into the ocean off southern California. Both were recovered with no loss of nuclear material.

Since its inception, the Soviet Union's nuclear program has had a cavalier attitude toward safety and the environment, for example in testing, production sites, power reactors, and submarines, so it should come as no surprise this carried over to its Space nuclear program. In 1969, two launches involving RTGs failed, both burning up upon re-entry and leaving detectable amounts of radioactivity. A 1973 launch involving a reactor failed, and it plunged into the Pacific Ocean north of Japan, releasing detectable amounts of radioactivity. In 1978, a reactor crashed into northern Canada, dispersing its radioactivity over a wide area. Canadian and U.S. crews picked up what they could find, but with the spring thaw, most of the radioactivity appeared to have sunk into the many lakes there. Since then, no detectable contamination has been found. Cleanup operations cost $14 million, with the Soviets paying $3 million. In 1982, another reactor failed to achieve orbit and returned to Earth, leaving a radioactive trail in the upper atmosphere over the South Atlantic. However, it isn't clear if any radioactive debris actually reached the Earth's surface. And in 1996, an RTG burned up somewhere over Bolivia and Chile or the southern Pacific, but no remains of the spacecraft have been found. In the 1980s, the Soviets changed their design philosophy to emphasize safety, so the 200 grams of plutonium-238 in the 1996 accident should have been contained and able to survive re-entry. Thus, when a design philosophy emphasizes safety, the use of atomic energy on land, sea, and in Space can occur without harm or undue risk, and I see no reason to assume it cannot be done for small and large engines.

airplane, and for chemically propelled rockets. However, I noted the Space Age has stagnated because of the insurmountable heat and weight limits of chemically propelled rockets, leaving Space expensive, elitist, and inequitable. We need a bigger hammer. Nuclear rockets are the bigger hammers, and they are limited only by our state of knowledge at a given time. Here I introduced the concept of the nuclear continuum, with the following conclusions on the develop-ability of nuclear rockets: Solid core: definite. Liquid and gas cores: maybe. Fission fragment: maybe, but doubtful. Fusion: not yet. Antimatter: iffy for a long time, if ever. I also noted the infrastructure for the nuclear rocket that developed during 1955-1973 was nonexistent now, and I held the law of supply and demand applies in Space as well as on Earth. In this chapter, I maintained the B-4 was the right choice of all solid core concepts, as it can achieve operational status the quickest. It will work. And it is inherently flexible, reliable, and amenable to increases in power density, specific impulse, fuel life, and other improvements. Then I introduced the small-large engine progression, and gave the technical specifications for fourth-generation systems, both as Re-use and Re-core models.

I will use these two in my argument to show how NASA can be restructured into an agency less dependent on federal funding yet with fundamentally new missions and supporting a vastly enlarged and privately funded space program. I start with the small engine in Chapter 4, go to the large engine in Chapter 5, and new NASA missions in Chapter 6. In Chapter 7, I discuss political infrastructure and in Chapter 8 I consider how nuclear rockets can achieve world peace or whether this is just wispy thinking by some of the world's best scientists. In Chapter 9, I offer concluding comments.

———————

At this point, I need to address the second 800-pound gorilla in the room: The fear the public will never accept nuclear rockets being tested on Earth because they discharge radioactive effluents that could cause irreparable harm to the environment. This fear is illusory, and simply not supported by the facts. In To the End of the Solar System, I included Appendix D, which looked at the safety and environmental record of the program over eighteen years, and concluded it was stellar. For example, nineteen of twenty reactors tested vented a radioactive plume directly to the environment, but different organizations, both reporting to and independent of NASA and the AEC, monitored it extensively. No public harm occurred. That is not just my conclusion, but also that of the EPA. It conducted a comprehensive review of the effluent releases a year after the program's complete demise – when it would have been easy to jump on NASA and AEC as "polluters" since SNPO did not exist and could not have rebutted the charge. However, EPA did not, but instead praised both agencies.

The EPA did not include the final reactor test, that of the Nuclear Furnace in June 1972, because it released virtually no radioactivity. It had a scrubber – in simple terms, it was several fire hoses that sprayed water onto the hot, radioactive plume coming out the nozzle. This cooled the plume, allowing the resulting mixture then to be collected and filtered to remove the

---

**Environmental Protection Agency Conclusion on Nuclear Rocket Effluents (1959-1970)**

"This paper has reviewed the history of nuclear rocket reactor engine tests at the Nuclear Rocket Development Station (NRDS), which adjoins the Nevada Test Site….It is concluded that off-site exposures or doses from nuclear rocket engine tests at NRDS have been below applicable guides. In general, it is felt that the program has been administered and conducted in a creditable manner and that the results reflect favorably on the management agencies."[1]

[1] D.E. Bernhardt, et.al., NRDS Nuclear Rocket Effluent Program, 1959-1970, (EPA, June 1974, NERV-LV-539-6, pp. 90-91

radioactive effluents and to discharge nearly pure gases to the environment. What remained was packaged for disposal. Scrubbers worked, and would eliminate the release to the environment of nearly all but trace amounts of radioactivity. (A trace amount does not mean it is harmful, only that our instruments are so sensitive that they can detect radioactive particles to parts per billion). In the mid 1960s, SNPO realized it could not continue to vent a radioactive plume to the atmosphere, but as the existing test facilities (all built in an era when atmospheric nuclear weapons tests were allowed) could not be easily retrofitted, it planned to move the program as swiftly as possible past them, to make them obsolete. In other words, they were *reactor* test facilities, but the program wanted to build *engine* and *stage* test facilities. So SNPO planned from 1965 on to add scrubbers to these new test facilities for the 5000 MW NERVA II and after its cancellation, the 1500 MW NERVA I. And Los Alamos planned a rail-mounted scrubber for

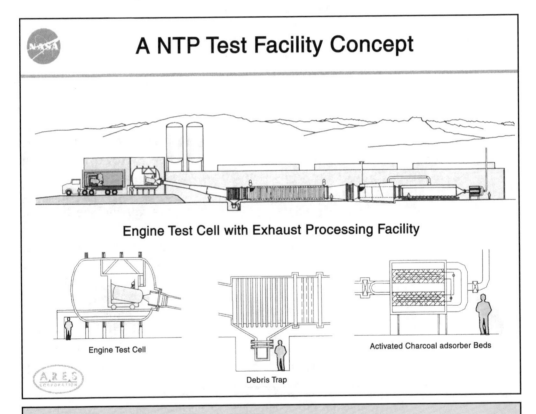

## A NTP Test Facility Concept

**Engine Test Cell with Exhaust Processing Facility**

Engine Test Cell

Debris Trap

Activated Charcoal adsorber Beds

CREDIT: MARSHALL SPACE FLIGHT CENTER

The Marshall Space Flight Center developed a scrubber concept similar to the one Los Alamos planned for the Pewee 3 test reactor, which was mounted horizontally on a rail car to fire into another rail car containing the scrubber. Basically, a scrubber would inject a water spray just below the nozzle to cool the super-hot gas, forming methane and other gases and absorbing the fission products. Then the mixture is cooled and filtered several times until it is clean enough to release to the environment and any residue packaged for disposal. In some designs, effluents such as hydrogen might be reused. One Russian nuclear rocket fuel test reactor had a scrubber, and both were located permanently inside a building.

the 500 MW Pewee 3 reactor. Yet some of today's nuclear rocket proponents as well as some in NASA and the space industry seem paralyzed by the fear of discharging radioactivity. They should reread the earlier scrubber literature.[v]

Instead, I hold the public will not be deterred by this illusory fear, and after they hear the arguments, will eagerly support a properly reconstituted nuclear rocket program, one that profits them personally, allows them personal access to Space and leads to an epochal change in world history. I maintain a great opportunity exists to convince them about this new democratic and fraternal space program. I am quite certain they will not support "mission-itis" thinking, particularly by those who aim it at expensive manned Mars missions. There is nothing in it for them except some photographs and a bigger tax bill.

---

[i] James T. Ramey, "The Development Process in Atomic Energy: The Dreamers, the Doers, and the Doubters," (remarks at the Chicago Chapter of the American Nuclear Society, May 22, 1963); Ramey, "Requirements Merry-Go-Round in Government Research and Development," speech given at American University on April 20, 1964 and reproduced as AEC S-8-64; Ramey, "The Requirements Merry-Go-Round: Must Need Precede Development," *Bulletin of the Atomic Scientists*, November 1964, pp. 12-15; Ramey, "The Requirements Merry-Go-Round – Phase II," remarks before the Panel on Space Capabilities and the Requirements Gap of the Atomic Industrial Forum, eleventh annual conference, San Francisco, CA, December 2, 1964.

[ii] Robert Seamans, NASA Interview, March 27, 1964, pp. 26-28; Seamans letter to Glenn T. Seaborg, December 12, 1962, DOE History Office, Box 1420, Folder 15, RD-29.

[iii] James E. Webb, NASA Authorization for Fiscal Year 1969, Senate Space Committee, April 1968, Part 3, p. 816.

[iv] Milton Klein to A.J. Eggers, "Potential Use of Nuclear Rockets in Second Stage of Launch Vehicles," December 8, 1966, NASA.

[v] "A total of six basic on-line scrubber system concepts have been evaluated, including the preparation of cost estimates. Four of the six concepts are directed at effective scrubbing of all engine effluent constituents except the noble gases krypton and xenon. The other two are intended to scrub all effluent constituents. Each of the first four systems will apparently provide effluent scrubbing efficiencies in excess of 95 percent for a NERVA employing composite fuel elements, and in excess of 99 percent for a NERVA engine with beaded fuel elements. The higher efficiency reflects the absence of the noble gases krypton and xenon in the source term for the beaded fuel elements. The estimated cost range of these systems is from $5,700,000 to $9,200,000. The latter two systems would be expected to give effluent scrubbing efficiencies in excess of 99.72 percent in all cases. They are both estimated to cost well over $40,000,000. In addition to the on-line systems, alternate effluent handling systems including off-line (storage-type) concepts were evaluated. Two basic schemes evolved, both of which would give efficiencies of better than 99.9 percent with either beaded fuel or composite. Costs in both cases would exceed $60,000,000. By allowing for the reuse of LH2 from these systems, and with a 50-test-per-year, ten-year program, the saving of approximately $30,000,000 in LH2 would put the effective cost of these facilities in the range of $30,000,000 to $40,000,000." Norman Engineering Company/Battelle Memorial Institute, "NERVA Effluent Scrubber Feasibility Study," Prepared for SNPO, August 17, 1970, p. I-2.

# Chapter 4
## If You Build It, We Will Launch It
### or
### The Free Launch

*This chapter considers how a small engine can <u>begin</u> to broaden domestic and foreign participation in the space program, increasing private, while decreasing government funding. Initially, five groups hitherto barred from Space will benefit from this wider access: Professional societies, amateur astronomers, students, inventors/entrepreneurs, and Greens, and as it increases, the expansion will allow citizens a return on investment of their tax dollars and expands the tax base. As this occurs, NASA's infrastructure can convert from an operational organization to one with private and public responsibilities, including supporting a burgeoning private sector space program. This will require legislation, including the creation of a private industry-government corporation.*

### Introduction

To discuss the above, I divide this chapter into four parts. The first outlines the technical infrastructure required to launch and recover Re-core engines from LEO, and Re-use ones for missions from LEO to points elsewhere. The second considers their costs. The third begins a discussion of the political infrastructure they require - the Democratization of Space Act - while the fourth considers two provisions of this act.

### I. Re-core and Re-use
### Engine Systems

Essentially, at least one Re-core and one Re-use engine would operate as a launch system. A Re-core would carry a Re-use engine, LH2 or payload to LEO, then return to Earth for a new core and then fly again, carrying payload or LH2 for the Re-use engine to

fly a mission. This would be mated to the Re-use engine, which then fires to carry the payload on its mission elsewhere, then returns to LEO for more LH2 and payload the Re-core had carried up. This would repeat until it has reached the end of its operational life of 30-hours with 180-recycles. This launch system has five components.

### A. The Re-core and Its Booster

The Re-core stage would be carried by a cargo plane such as the C-5A to about 10 miles (50,000-feet), and then dropped, whereupon solid strap-on booster rockets would fire and carry the stage to more than 100,000-feet or over 20-miles high.[i] Then the Re-core would fire to carry the payload to LEO, typically somewhere around 200-miles up. The stage would be comprised of the Re-core at ~6000-pounds, a 17,000-pound payload and a 3000-pound cocoon (below). I assume the

LH2 tank would weigh ~20,000-pounds and hold 90,000-gallons of LH2 that would weigh ~45,000-pounds, enough for a 15-minute flight to LEO, assuming the turbopump uses 100-gallons per second (This is a generous amount since the flight would be closer to 10-minutes, but these assumptions suffice as I am not trying to design a nuclear stage, but present a new way of thinking about them). The total would be 91,000-pounds, plus the weight of the strap-on solid rocket boosters.

---

### SPECIFICATIONS OF FOURTH-GENERATION RE-CORE ENGINE AND STAGE

**Engine**
800 MW (40,000-pounds of thrust)
6000-pounds engine weight
1000-seconds of specific impulse
100-gallons of LH2/second

**Stage**
LH2 tank weight: 20,000-pounds
17,000-pound payload
3000-pound cocoon
LH2: 90,000-gallons/45,000-pounds

**Totals:**

| | |
|---|---|
| Engine | 6000-pounds |
| LH2 tank | 20,000-pounds |
| LH2 | 45,000-pounds |
| Cocoon | 3000-pounds |
| Payload | 17,000-pounds |

91,000-pounds

---

I have four reasons for this launch sequence. <u>First</u>, it permits the Just-in-Time launch whereby the Re-core, the LH2 tank, the cargo plane, and payload were kept separate until the latter was ready. Just-in-time manufacturing means keeping inventory levels to the barest minimum necessary to keep the production line operating. This minimizes the money tied up in idle parts sitting on a shelf, thus lowering costs and improving profitability. When it was, all four would be taken to an airport and assembled, thus eliminating security risks, as the engine would be kept under DOE control until the flight was imminent. In addition, it ends the need for permanent installations, such as the Kennedy Space Center and its enormous and costly infrastructure. Almost any airport could be configured to handle mating the Re-core, LH2 tank, and payload, and other than needing the ability to handle LH2, it needs special security only when the nuclear engine arrived. So costs here could be minimal and fixed installations could be rededicated to supporting a private sector space program.

Second, the cargo plane launch allows the Re-core stage to be flown to an isolated ocean area where the flight profile would be over water, ending any real concern of a nuclear rocket crashing back to a populated area. This represents little more than applying the zoning concept to isolated ocean areas

that we use for our cities; e.g., some zones are residential, some commercial, some for light industry, and others for heavy industry where there are risks of accidents. All fixed launch complexes use it and have launch corridors to minimize the risks to the public from errant rockets. Now, however, isolated areas thousands of miles from population centers would be so designated. The cargo plane would act as a rocket's first stage, and this is a well-known technique to save on launch costs. It is similar to DARPA's RASCAL.

---

**DARPA RASCAL**

CREDIT: DARPA

Launching a rocket into LEO via an airplane is a long-established concept. The most recent version was DARPA's RASCAL, a supersonic airplane that was to carry a rocket on its back to 100,000-feet, where it would be launched to carry its payload into LEO while the plane returned to its base. The rocket would not be recovered. The program was canceled after spending $200 million because the high-tech supersonic aircraft became more expensive to develop than anticipated. The flight profile for Re-core engines would be similar to RASCAL's, but the plane itself would be a cargo plane such as C-5A fitted to carry the stage. This is existing and proven technology, so such a plane should not be expensive to procure.

---

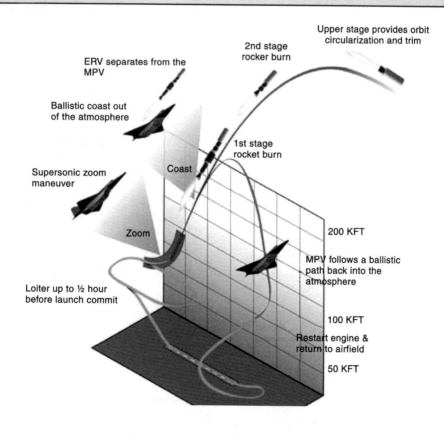

Upper stage provides orbit circularization and trim

2nd stage rocker burn

ERV separates from the MPV

Ballistic coast out of the atmosphere

1st stage rocket burn

Supersonic zoom maneuver

Coast

Zoom

200 KFT

MPV follows a ballistic path back into the atmosphere

Loiter up to ½ hour before launch commit

100 KFT

Restart engine & return to airfield

50 KFT

Third, once dropped, solid strap-on rockets would boost the stage to more than 100,000-feet, increasing the stage's overall speed, but more importantly, allowing the Re-core to startup and come to full power. This takes time, and allows the cargo plane to fly a safe distance away from any radiation. Remember air absorbs gamma rays and neutrons in three-fifths of a mile at sea level, but that distance increases at higher elevations where the atmosphere is thinner. And since the Re-core would reach full power somewhere above 100,000-feet, about where the vacuum of Space begins, this allows the use of a more efficient rocket nozzle designed for space oper-

ations. After separating from the stage, the cargo plane returns to an airport.

Fourth, the cargo plane would obviously be used repeatedly, thus lowering costs. However, it isn't clear whether the C-5A could really launch the Re-core stage. It certainly has the lifting capacity and the stage could fit, though snugly, into its large interior, but whether it could be drogue-parachuted out of the tail of the C-5A is uncertain. Undoubtedly, if the C-5A could be used, it would need modifications; but if it couldn't, the best option is to design a cargo plane just to carry the Re-core stage.

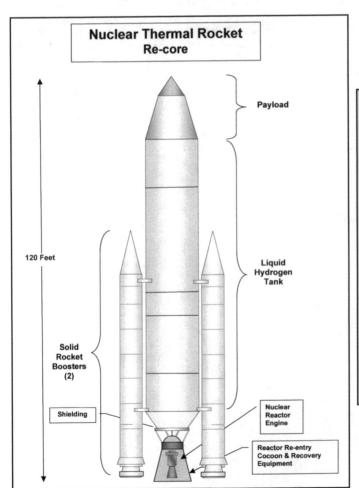

**Nuclear Thermal Rocket Re-core**

Payload

120 Feet

Liquid Hydrogen Tank

Solid Rocket Boosters (2)

Shielding

Nuclear Reactor Engine

Reactor Re-entry Cocoon & Recovery Equipment

**RE-CORE STAGE**

The Re-core stage as depicted should fit snugly inside the C-5A's cargo bay, but if for some reason it could not carry it, another cargo plane must be developed to do so. In this illustration, the nuclear engine is pictured as being enclosed in a re-entry vehicle type cocoon, similar to the Mk-6 for the Titan ICBM. It permits the engine to return for recovery without undue risk. Appendix C will illustrate other possible cocoon shapes.

At this point, the reader may want to refer to Appendix C to see the Re-core's flight profile explained in a non-technical way to address the fear of a nuclear engine crashing during launch, spewing radioactivity going to LEO, or disintegrating in re-entry. My ten-step analysis illustrates how it could be done, and a demonstration program, I believe, would prove the risks would equal those associated with the airline industry. And with experience, I believe those would become even less than those we accept for flying.

## B. Recovery of the Re-core Engine

Now space operations begin. The Re-core must be separated from its empty tank, a relatively easy task, before its return to Earth, but to do that it must be protected from the rigors of reentry and its tremendous heat and shakes, rattles and rolls. Returning a radioactive reactor to Earth *intact* is seldom studied and in the only study I am aware of, the author would place a heat-resistant shield atop the engine and a plug in the nozzle to prevent sea-water from entering the core.[ii] This idea leaves the pressure vessel as the main line of defense against accidents and the rigors of re-entry. A much better approach is a cocoon to envelop or surround the engine, one shaped like the Mk-6 re-entry vehicle for the W-53 warhead, a blunt mushroom Orion/Apollo-type shield, or a lifting body. These are proven but when linked to a nuclear rocket, they are totally new and never studied; the Re-core's 20,000-pound payload would

allow considerable freedom in design-ing one, so it's difficult to say what shape would be best. Moreover, a clamshell shield might be devised, similar to that for the XE engine test in 1969 that enclosed the engine. Here the clamshells would be open for the flight to LEO, but close for the return to Earth. Whatever the choice, a buoyant cocoon with the capacity for a pinpoint landing and soft touchdown and the ability to carry accident mitigation or prevention gear would be key design and performance objectives. Whatever its design, it would undoubtedly con-tain poison materials such as boron to absorb much of the radiation. For my argument, I'll assume it would weigh 3000-pounds; this leaves a net payload of 17,000-pounds.

On returning, the cocoon must be recovered, and recovering items from orbit has occurred repeatedly since the beginning of the Space Age, on land, at sea, even in the air. The fact that the cocoon contains a radioactive engine would be new but would not pose a problem. Why? The federal govern-ment has extensive experience dealing with nuclear contingencies that would be readily adaptable to this new mis-sion. The military has its Broken Arrow/Broken Shield capabilities to respond to an accident or loss of a nuclear weapon, DOE has its Nuclear Emergency Support Team and safe/secure transportation system, and the Nuclear Regulatory Commission (NRC) has its emergency capabilities for nuclear sites in the public sector. And NASA has some capabilities, derived from its launching nuclear power sources. All this is not quite what would be needed, so this infrastructure must be developed; these capabilities, however, should be easy to adapt to create it. Also, I expect the cocoon would land at sea via parachute initially, perhaps near a U.S.-controlled island in the Pacific, so the navy would be involved. Finally, a cask must be developed to hold the cocoon, as it would still be emitting radiation, though the cocoon would dampen much of it. This is a routine task. As experience is gained, the Mk-6-like or Orion/Apollo-like cocoon might give way to the lifting body, and a pinpoint landing on an isolated land area, perhaps eliminating the need for the navy's presence.

### C. Inspection, Servicing, and Recertification: Re-core Engines

Now the cocoon must be transported to an area where it and the pressure vessel would be opened, and the radioactive core removed and sent for recycling to recover the uranium. Then the turbine/turbopump and other non-nuclear components would be re-certified, rebuilt or replaced. Afterwards, a new, fresh core would be inserted, and the engine readied for flight. These functions would be conducted at Jackass Flats or other DOE sites where hot cells exist to remove the core, but rebuilding and recertifying the other components and inserting of a fresh core would not need hot cells or remote handling equipment. This could be done manually, with minimal radioactive safety requirements. A re-certified Re-core would then be ready for another Just-in-Time-launch.

### D. Re-core Re-launch

With its new core, the Re-core would await its payload and LH2 tank and once available, mated to the cargo plane; then the launch sequence would commence again.

### E. Space Operations Infrastructure: The Re-use Engine

Once the Re-core returns to LEO, it must be de-mated from its payload, which in turn must be mated to the Re-use engine that has been waiting there. Then it would fire, carrying its payload to points beyond LEO, and once completed, fires again and returns to LEO, to be inspected and re-certified again as mission ready. Rover/NERVA gave this mission profile considerable attention, and it would face some problems outlined in Appendix C. I will not consider them in this book. I can say that technically some of the work would involve inspecting the core (via cameras and non-destructive testing) to re-certify it, servicing/replacing non-nuclear components such as the turbopumps and mating new cargos and refilling the LH2 tank. Robots could carry out these functions initially, but later, hands-on maintenance would be likely, with the Re-use engine enclosed in a heavily shielded structure that houses master/slave manipulators, a sort of LEO "garage." Here astronaut "grease monkeys" would peer at the engine through thick lead glass windows and perform their activities with a direct line of sight. In essence, this "garage" would be a hot cell common to the nuclear industry, but now in LEO. It would be heavy so I discuss it in the next chapter.

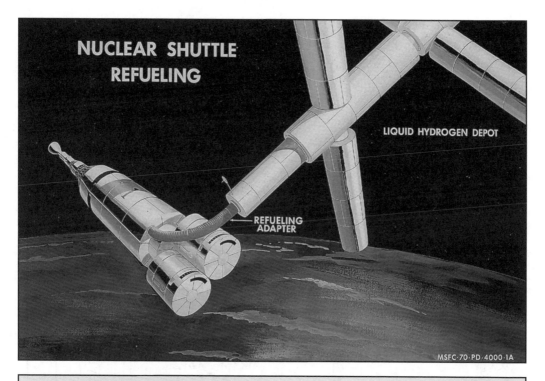

**NUCLEAR SHUTTLE REFUELLING**
CREDIT: NASA
NASA studied refueling NERVA engines extensively in the 1960s for many deep space missions, including manned Mars. These studies form points of departure for the "garage" for small Re-core and Re-use engines.

Before proceeding, I will recap the technical infrastructure. It consists of the following components:

- cargo launch plane
- cocoon and recovery infrastructure
- Re-core engine development and re-certification infrastructure
- airport launch infrastructure
- space operations and Re-use engine infrastructure (the "garage")

## II. Economic Costs

I must now consider the economic costs for this, leaving aside for the moment where the money comes from. In 1972, the cost for developing the Small Engine for the space shuttle through the flight test was $500 million. That estimate would be fairly accurate for a Pewee-type system, but must be normalized into today's dollars, so I'll say $2 billion. To this must be added money to refurbish the decayed test complex at Jackass Flats or elsewhere, and I'll assume this to be $2 billion. Next, I assume the cargo plane launch and recovery infrastructure would cost $2 billion. So now we are at $6 billion. Now I'll assume each fourth-generation Re-core would cost $25 million and its non-core components would have an operational life of 100 flights, but that each new core would cost $1.6 million. I obtain that cost as follows: In the 1960s a fuel element cost $1000 to make so I assume that would be $4000 in today's dollars.[iii] Each Re-core would need 400 fuel elements, hence the $1.6 million for each core and $160 million for 100 flights. The total, so far, is about $6.16 billion, and for this amount a fourth-generation Re-core could propel 1.7-million pounds to LEO in 100 flights.

That's about $3600 per pound if all development costs were amortized into the fourth-generation engine. However, in the real world, costs are not be amortized this way, so if I just amortize the costs of the $25 million Re-core and its $160 million for 100 cores, that's $185 million. So for that amount, 1.7-million pounds could be taken to LEO for about $108 per pound.

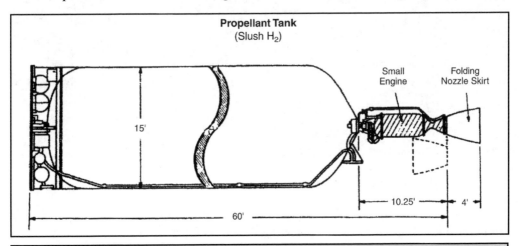

**Propellant Tank**
(Slush H$_2$)

Small Engine

Folding Nozzle Skirt

15'

60'

10.25'

4'

**THE SMALL ENGINE**

CREDIT: LOS ALAMOS NATIONAL LABORATORY

In Rover/NERVA, the primary emphasis was on large engines for manned missions, but smaller ones were studied over the years. This small engine of 1972 was to be carried in the cargo bay of the space shuttle whereupon it would operate from LEO until it had reached its 10-hour/60-recycle fuel element lifetime. Los Alamos estimated its cost would be $500 million in 1970s dollars for a first-generation system.

Figuring Re-use engine costs are quite complicated, but for my argument I'll assume two scenarios. In the first, the engine would fire for 30-minutes to leave LEO and 30-minutes to return. Since it would have 30-hour fuel, this means 30-round-trip missions, each propelling a 20,000-pound payload. (I assume the Re-use engine would not have a cocoon). In the second, I assume an hour firing out and back, so there would be only 15 round-trip missions, now propelling only

300,000-pounds. Since a Re-use engine and its fuel would be more costly, I assume each would cost $100 million and so, for that amount, would be able to propel 600,000-pounds beyond LEO for $166 per pound or 300,000-pounds for $332 per pound. This is clearly wrong, so don't take these numbers as gospel. Developing good data will take time, so this gives a rough insight into the costs and they probably won't be much.

---

### Comparison with Chemically Propelled Rockets

The space shuttle can take a pound of cargo to LEO for $15,000 while the new Delta 4 Heavy, at a cost of $140 million each, will take 48,000-pounds to LEO. That projects to $2900 per pound.

---

In sum, a Re-core could carry 1.7-million pounds to LEO in 100 flights for a total cost of $6.16 billion or about $3600 per pound if all its development costs are amortized. If not, the Re-core's costs could drop to $108 per pound. The Re-use engine could carry 600,000-pounds beyond LEO for $166 per pound or 300,000 for $332 per pound.

Are these figures accurate and reliable? Absolutely not! They are hypothetical and derive from my conception of a flight profile NASA has deliberately prohibited for half a century, from an engine concept – the Re-core – no one has ever studied, from an overly generous 15-minute flight to LEO instead of a more realistic 10-minute one, and from a cargo plane that permits the Just-in-Time launch. So everything is new. Also, private sector accounting would never permit so simple a method of determining costs; much more sophisticated methods would prevail. I'll discuss a third reason shortly. So these figures are certainly wrong. A hundred-plus dollars may be too low. On the other hand, if some of the "Wild Card" improve-

ments discussed in Appendix B are realized in a fourth-generation Re-core, a hundred-plus dollars may be too high. Whether too high or too low does not matter for my argument, as these figures are reasonable conjectures based on reasonable extrapolations of improvements to the B-4 core, ones realizable in a decade. So they are accurate and reliable enough to show extraordinary improvements to the economics of the space program. They are accurate and reliable enough to justify breaking the taboo against using nuclear engines to reach and return from LEO, and repudiating the "mission-itis" attitude that so afflicts the space program. They are accurate and reliable enough to justify adopting an approach that first develops prototypes and infrastructures, with missions evolving from that. And most important, they are accurate and reliable enough to consider the policy implications of such systems. These will be shadowy and vague now, like seeing shapes through a frosted window, but there will be winners and losers, there will be consequences. I discuss them next.

## The Nuclear Fleet Model

Developing reliable figures on Re-core and Re-use engines requires a different way of thinking than with chemical rockets; that statement includes my most cursory analysis. Except for the space shuttle, here the thinking is one rocket, one mission, with months or years between the fabrication of a new rocket and its payload and its actual launch, a time in which highly trained launch and operations personnel are paid for ancillary work. In contrast, my analysis reveals the increased payload and reusability of nuclear rockets will reduce costs dramatically, causing an equally dramatic increase in demand. Thus, to use the chemical rocket model with its months or years between flights is inappropriate. Different models and different thinking are needed.

I hold the experience of the competitive airline, shipping, train, and trucking sectors, honed to a fine edge to stay profitable, should be mined to see how it might be adapted to nuclear rockets. To dismiss them as unrelated to Space is shortsighted, as they operate fleets traveling between different destinations. So too will nuclear rockets. Thus, similarities exist. Even the original space shuttle model, with flights every month, might provide insight. Some may disagree and point to the shuttle's high costs and lengthy time between flights. That compares apples and oranges. Perhaps the basic reason the space shuttle failed is the unreliability of its delicate LH2/LOX engines, a figure NASA does not publicize. In contrast, nuclear engines have a simple ruggedness, a 99.7 percent hardiness, making them similar to the robust and reliable ones powering airplanes, ships, trucks, and trains. If these industries had to operate with engines as unreliable as the shuttle's LH2/LOX one, they would be bankrupt quickly.

Thus, a fleet model must be developed to give precise economic figures *and* become a key part of the research and development process. Rover/NERVA used modeling extensively, with codes developed to predict reactor/engine operation with remarkable accuracy, given the 1960s-era computers and codes it used. Likewise, TRW computer studies in the mid-1960s demonstrated the interrelationship of missions to engine design. However, neither was linked to economics and infrastructure. NASA always planned to use a nuclear rocket from LEO, with the Saturn or shuttle taking it there, all with public money, not private. So new models must do what the older ones did, but also show the interrelationship of engine development, infrastructure, and economics, i.e., the savings in transitioning to later-generation engines, and they must also point out specific areas for engine and infrastructure development to affect those savings. The cost of keeping spare parts on hand must be included, and doing this requires probabilistic studies on the life expectancy of turbines and turbopumps, pressure vessels, nozzles, and so forth.

In sum, I expect nuclear fleets to be built, based on fourth-generation Re-core and Re-use engines, but can say little beyond that except that each fleet might have ten rockets: One being launched to LEO, one departing and another returning from LEO, one cooling down in LEO, another returning to Earth, one in re-inspection and re-certification, another in the launch sequence and several in reserve, ready to respond to the market. An ample supply of parts would be needed to keep all running on a schedule. If I am correct, it already reveals economies of scale, as building fleets of ten or more would cost much less than building one at a time, as is done now. These things would be defined more precisely in building the small engine generations one, two, and three.

## III. The Democratization of Space Act

I begin by considering the reaction of domestic and foreign firms and their governments once the U.S. Congress breaks the taboo and announces the four waves of small-large engine development. I am certain they will know instantly their rocket programs are in deep trouble. They know the rocket equations and heat and weight, they know that increasing specific impulse means reducing mass ratios while increasing payload fractions, they know this intuitively without knowing a single specification for the size and shape of a nuclear engine, and they know when this happens their rockets will become obsolete. To drive home this point, I'll be blunt: They will be history, goners. The only question is when, and that will be soon. The ponies of the Space Age – the Ariane, Atlas, Delta H-2A, Long March, and Titan to name a few – will be put out to pasture. This includes the newer vehicles such as the Air Force's Evolved Expendable Launch Vehicle and NASA's Ares. Used once then thrown away, they simply cannot compete with $108 per pound to LEO and $166/332 beyond it. This is true even of countries such as China with its low rocket production and launch costs. Hence, expecting fierce opposition from them is logical, as they fear the loss of business and competitive positions – indeed, their very existence. That may happen, but it need not because startling consequences lay just below the surface. The why is simple.

In the U.S. political and economic system, the government cannot compete with the private sector. This means if the government developed Re-core and Re-use engines directly or indirectly, once they have been demonstrated and an infrastructure debugged, it must divest itself of them. Now three parts of the infrastructure could be privatized: Launch, cocoon recovery, and space operations. However, since an engine would use HEU, a government license would be required, with oversight as appropriate. Moreover, all development must be conducted at government facilities for security reasons, but launches could be from public or government airports. Just bring in security when needed. Recovery operations would be conducted at sea initially, then perhaps in isolated areas under government control later.

Developing new engines and re-certifying used ones, however, is a complex issue, as a close connection will exist between engine design/manufacture and the government's role to pioneer new ideas and concepts. I must consider this in detail. Rover/NERVA demonstrated the close relationship between the pioneers (Los Alamos) and pragmatists (Westinghouse and Aerojet General). It will exist here also, as flight experience will influence pioneers and pragmatists and result in later generations of engines with vastly improved capabilities. The government's pioneering role will stay the same and be like DOE's relationship to the nuclear (and NASA's to the airline) industry. Both conduct pioneering and safety research, but do not infringe on the marketplace.

This leaves defining the pragmatist's role, and for this I see four options. The most obvious is to use the 1960s model where one or two firms (Westinghouse and Aerojet) produce first, second, even third-generation systems under a NASA/DOE contract. However, at some point a transition must occur, most likely with a fourth-generation system. Afterwards, later-generation systems would be produced for the government and private sector, optimized for different missions. These would have no direct government funding. Now if only one or two firms exist to do this, then few if any competitive problems would exist in using data gained from flight experiences. That solves one problem, but creates an insurmountable one. If a single vendor with a sweetheart government contract produces nuclear engines capable of carrying payloads around Space for $100 a pound, it privileges one firm while obsolescing all others. That is unrealistic. I cannot see the United States withstanding the domestic and foreign criticism and pressure if this happens. So this 1960s model is untenable ultimately.

The problem then is how to structure an engine development and operations program that has a role for the pioneers and allows for the production of successively improved engines while at the same time using flight data to improve all **and** using this data without compromising the competi-

tiveness of the different firms **and** doing all in a way that provides a return on investment to taxpayers. This situation is complicated even further by security concerns, since the engines would use HEU. Domestically, that is not a problem, but it is for international participation. Yet the foreign outcry would be loud and strident if other nations were excluded.

A second option to consider is the Power Reactor Demonstration Programs of the 1950s in which the AEC, nuclear vendors, and utilities cooperated in developing the light water reactor. Once demonstrated, utilities now had competing vendors from which to choose. Under this approach, different engine manufacturers would participate in developing first-second-and-third-generation engines, perhaps under a cost-sharing arrangement with NASA/DOE, but be on their own for later-generation systems. If agreements for cooperation were in place under the Atomic Energy Act (which governs reactors), each company could enter into arrangements overseas. This would predetermine government approvals and allow individual companies to decide the nature and extent of their partnerships. This model proved highly successful, as U.S.-origin light water reactors are dominant in the world. However, it may not solve the problem of sharing flight data or allow any return on investment for taxpayers, and it raises the specter of having to get NRC export and import licenses for each flight. It seems cumbersome and unworkable, and may require a lot of up-front government funding.

A third option might be Comsat/Intelsat, the multinational entity created in the 1960s to allow international participation in communication satellites. It has worked well, but is centered on satellites, not developing rocket engines and launch and missions operations, and it provides a return only to stockholders, not taxpayers. I don't think it is workable.

The best approach, I hold, is to think through the entire matter afresh as it involves so many new and different factors. This will require new legislation, as the laws creating NASA and DOE were written decades ago for different purposes. So I propose the Democratization of Space Act. It will have many sections, some of which I list in the boxed text but will not discuss in this book (they're important but not to my argument's main thrust) and some of which I will discuss later.

---

### The Democratization of Space Act

* Creation of Nuclear Rocket Development and Operation Corporation
* Privatization of cocoon recovery/emergency response infrastructure under U.S. license
* Privatization of launch and re-launch infrastructure under US license
* Creation of Assistant Secretary of State for Space Nuclear and Nonproliferation
* Creation of NASA Office of Public Missions
* Creation of NASA Office of Space Station Design
* Provision for EPA and public environmental monitoring and waste disposal programs
* Provisions for foreign participation in engine development and operations
* Provisions for domestic and foreign public participation and remuneration
* Provisions for transparency and openness consistent with protections for intellectual property and commercially valuable property
* Provisions to expand Patent Office to handle increased workload

---

Here, however, I will discuss the section that creates a Nuclear Rocket Development and Operations Corporation (NucRocCorp). I cringe at this awful bureaucratic name, but since I am only presenting a concept for consideration, it will suffice. NucRocCorp would merge DOE and NASA assets and their pioneering role with those of the rocket engine firms, the pragmatists, creating a for-profit COGO – corporation owned, government oversight organization. This is a variation of NASA's GOGO laboratory structure (government owned, government oversight) and DOE's GOCO laboratory structure (government owned, contractor operated). Here the United States would transfer to NucRocCorp the rights to use the land, buildings/facilities, and infrastructure at Jackass Flats and other relevant DOE and NASA sites for a nominal fee, such as a dollar a year, and DOE would provide the uranium and security at no cost. All this would be a contribution in kind. This government freebee is the third unknown in determining the costs of taking payloads into Space and, depending on how it is structured, could be a critical factor in reducing costs. It would ease security concerns because DOE systems would be in effect - its guards, gates and guns and safe/secure transport system - and it would retain ownership of the HEU. Meanwhile, the U.S. and foreign governments that join NucRocCorp would contribute *some* money, *if necessary,*

one of three sources of funds, to develop the first-through-third-generation systems. Afterwards, however, this government contribution would decrease for engine development and shift to funding space infrastructure, *if necessary.* Then it would decrease again and shift to fund safety and other research, like what DOE and NASA do for the nuclear and aerospace industries. I've emphasized the words *some* and *if necessary* by italicizing them as I wish to stress to the reader that two other sources of funds are available that may eliminate the need for government funds totally. I discuss them shortly. But, if they are necessary, they must be repaid. There will be no permanent line item in the budget for them.

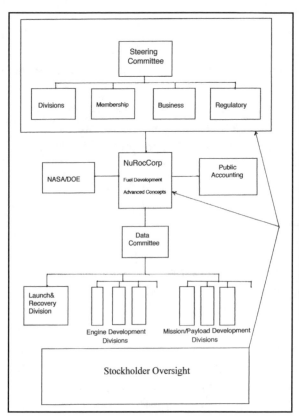

**NUCROCCORP ORGANIZATION CHART**
I introduce the NucRocCorp concept here but will explain the roles of the various groups further in chapters 7 and 8. This concept includes boards of directors supervised by a Steering Committee that will be transformed into different entities when large engines appear. Their role with the smaller engines is to prepare for that future.

Rocket engine firms would transfer their assets (people, hardware, and cash) to NucRocCorp, to complete the infrastructure at Jackass Flats or elsewhere for engine development, production and operation. To do this, they must pay cash entry fee, the second source of funding, and it should be substantial to eliminate the non-serious or those on fishing expeditions. It might be $250 million, $500 million or even higher per firm. In other words, let me be blunt: This alone could be a big pot of money, and cover the $6 billion figure I mentioned; if so, no government money would be needed. Despite the large price tag, rocket firms have only two choices: Come up with the cash or go out of business. I think they would see the first as a no-brainer, particularly when they see NucRocCorp's extremely large profit potential and its friendly and inviting structure.

It might be similar to General Motors, which has overall corporate governance with semi-independent, competing yet cooperating divisions such as Cadillac, Buick, Pontiac, and Chevrolet. Each firm joining NucRoc-Corp could establish its own division or enter into partnerships to form one. I see three separate groups being created: One or more engine development divisions and one or more payload development and mission divisions. I expect there would be a single launch and recovery division initially. Each division would select officers to manage NucRocCorp, while the United States and other governments would pick their representatives for the boards of directors for oversight. This is plural and I will explain why in chapter 7. Each division would compete in the marketplace under rules that allow it to keep a preponderant share of its profits. The other share would fund the NucRocCorp structure and provide a return on investment to its stockholders. Perhaps NucRocCorp's most striking feature would be that for-profit managers would be in charge throughout the organization, not bureaucrats, yet the United States and other governments would have a say on policy via their membership in a Steering Committee. Finally, as government pioneers would work for NucRocCorp (in fuel development and advanced concepts), it would own any patents, intellectual property, and know-how they develop and market these as appropriate. Moreover, each division would retain rights to any valuable items its employees make. I expect all this to be highly profitable. To say it would not happen ignores the technologies deriving from Rover/NERVA, such as the multi-billion-dollar graphite and heat pipe industries.

## The Quid Pro Quo

Governments joining NucRocCorp would have an obvious quid pro quo for so doing. They must be able to purchase or lease engines or launch services for governmental use at cost – not market-rate, not discounted, not cost plus fixed fee or cost with fees buried somewhere, but cost. Conditions would have to apply obviously, foremost of which would be that a government's use of such an engine must be for legitimate government purposes, ones that clearly do not infringe on the marketplace. Also, governments should have access to the books to determine the actual costs.

Some may sneer that's only a one-time example from the past, so I'll give two other examples. NucRocCorp would oversee a major high-temperature materials research and development effort, and as these materials appear for nuclear engines, they have widespread private sector uses. Materials operating at ever-higher temperatures promise higher efficiencies that promise greater savings and profits. So I expect extensive private sector demand for this. Moreover, as noted in my preface, new compact high-temperature gas core reactors could be developed for producing electricity as well as for process heat, ones with low capital costs to build and make secure. This could be worth billions, particularly ones to melt wastes. So I would expect NucRocCorp and its divisions to charge what the market would bear to maximize a return for its stockholders.

I think this concept could withstand the further scrutiny it requires, as it has many advantages. It provides a role for the U.S. and other governments, but reduces or eliminates the disruptive effects of the budget process on engine development and space operations. Private money would flow into NucRocCorp, and it would expect for-profit managers to use it wisely and provide a return on investment. It allows for flight data to be shared and for innovations by the pioneers, yet it also allows for competition among divisions. To believe this would not take place, since only Re-core and Re-use engines would exist, is dead wrong; many variations would come

into being, particularly as engines were optimized, and as competition took root. It permits each division not only to offer launch services, but also allows the parent firm to provide the payload for the customer, as many rocket engine manufacturers already do. This will be, I submit, highly profitable, as it is premised on the law of supply and demand; as the cost of moving payloads into and beyond LEO drops significantly, not only will the number of missions increase but also their costs will decrease, too. Indeed, I expect these payload/mission divisions to experience dynamic growth, as the private sector takes advantage of ever-lower costs. Likewise, it has three options for foreign participation: One on missions whereby foreign entities might obtain via a lease Re-core or Re-use engines for their own operations. Or they could partner with the existing divisions under terms and conditions they negotiate. Or they could establish their own divisions for engine development, payload/missions and launch and recovery within NucRocCorp. They just need to pay the entrance fee. Some countries might be barred because of nonproliferation and other concerns. Canada, Mexico, the European nations, and Japan would pose little risk here, but others must be handled on a case-by-case basis. I discuss membership in chapter 8.

Finally, NucRocCorp will allow taxpayers to participate and get a return on investment. This is the third source of funding, and it could be another large pot of money. How could that happen? The Democratization of

Space Act must require each 1040 tax form (other forms for other countries) to contain a line that allows each citizen to check if he or she wants some of their tax refund to purchase shares in NucRocCorp. The 1040 form is appropriate, as John Q. Public's taxes originally funded the DOE/NASA infrastructure, so when part of it is restructured into a for-profit company, he or she must have the right to invest via the same form. This would be neither new nor novel, as Congress has used the 1040 form to stimulate the public's interest in many topics, e.g., tax credits for solar energy. The price of each share should be modest, say $50, and each taxpayer must purchase one share in NucRocCorp. Afterwards, they could purchase as many shares as they wanted in NucRocCorp, any division, or fund any mission. This extra money could come from the tax refund or personal income or it could be venture capital money. It would be voluntary, there's no coercion, and would establish a simple and easy way so all taxpayers could participate, yet it allows for further participation.

That could be extraordinarily significant. For example, if only 10 million U.S. taxpayers purchased a single $50 share of stock a year, the NucRocCorp would have an infusion of $500 million annually; if 20 million, then $1 billion. If the same number purchased a $10 share in a division, it would have $100 million. If the same number paid just $1 to fund a particular mission, it would have $10 million. Thus, a large, untapped source of *interest-free* money could exist to further engine develop-

ment or conduct missions while at the same time reducing the government's obligation. For the United States, it ultimately might be just a contribution in kind, with no outlays of money. And stockholders would have oversight of NucRocCorp and boards to ensure their money is used to give a return.

Then we add foreign participation. Citizens of member countries would have the same rights of access within NucRocCorp as U.S. taxpayers. Their governments would find membership highly attractive, as it probably involves no outlays of money, yet private sector space industries grow in their country, broadening the tax base. As these governments would have oversight via the Steering Committee, this helps ensure they and their citizens receive fair treatment and access. Yet the Steering Committee/boards structure allows NucRocCorp or its divisions to pursue commercial arrangements overseas knowing they would be pre-approved or subject to quick approval. All this could transform foreign opposition to launching and recovering nuclear engines into active support, as they have the right of full participation. This would dramatically enlarge the pool of investors and the *interest-free* capital NucRocCorp and its divisions would have. At $50 a share, it could become a huge worldwide pot of money. If I may paraphrase Everett Dirksen, a Republican senator from Illinois in the mid-twentieth century, a billion here and a billion there and soon you have some serious money. Finally, offering low cost launch services overseas would give diplomatic leverage over states with questionable missile and nuclear programs. They would be harder to justify as peaceful when lower cost alternatives exist. I discuss this in chapters 6 and 8.

Hereafter, I assume the NucRoc-Corp model has merit, but now will consider the service bay or "garage" in LEO. NucRocCorp should fund its development and launch, since its design will follow that of engines, and manage its operations, since it would train the astronaut "grease monkeys" to service them. Over time, more than one "garage" would be required as the volume of activity increases. Later, after serving their demonstration purpose, NucRocCorp should sell them; henceforth, they would charge a fee for service.

Let me now pause to summarize my argument. I assume Congress passes a Democratization of Space Act creating NucRocCorp, a for-profit, contractor-owned, government-oversight corporation to develop, launch, recover, and operate Re-core and Re-use engines. It would have a privately operated infrastructure, under government oversight, and allow for domestic, foreign, public, and private participation. I have assumed a cargo plane would launch a fourth-generation Re-core engine at 50,000-feet where it would propel a 20,000-pound payload into LEO, minus a 3000-pound cocoon, for a net of 17,000-pounds per flight, 1.7-million pounds for a hundred flights, at a cost of $108 per pound. A Re-use engine would make 30 or 15 flights during its lifetime, propelling 600,000/300,000-pounds beyond LEO for maybe $166/$332 per pound.

## IV. Two Key Provision of the Democratization of Space Act
## A. The Free Launch

Here I consider two key provisions of the Act, and to do this I must focus on the 1.7-million pounds. Now my argument becomes complicated, as so many missions exist. I start with the simplest and assume the 1.7-million pounds would only go to LEO. A Re-use engine would not carry it elsewhere, but even this simple scenario becomes complex. How would this total be allocated? If all for science, that would please the scientists immensely as it could mitigate, perhaps eliminate, the friction of selection - of one scientist winning the right to have his or her experiment launched, and another losing. So large a total means all experiments could be taken at least to LEO. Moreover, they could be more frequent, so scientists would not have to wait years to obtain their data. Henceforth, there might be only a queue, and perhaps a rapidly moving one at that, and one requiring a lot less paperwork. Finally, the selection process could change. The traditional criteria of scientific importance, contribution to knowledge, technical feasibility, competence of the experimenter, and need to fly could lessen; at the same time, avoiding interference among experiments, handling the telemetry of increased amounts of data, and providing power to all experiments could grow in importance.[iv] Here payload specialists might come to the forefront, with scientists receding in the selection process, with the payload ultimately becoming similar to an airplane with a number of seats. Each scientist would be given a seat – or so many cubic feet and so much electricity for his or her experiment – and when the plane was full, it took off. The Re-core's 17,000-pound payload would allow substantial latitude in the design of these "airplane seats," perhaps with cost-saving standardization the end result.

I expect NASA's Office of Space Science would administer the program though now it would be more democ-ratic, and I anticipate a broad con-stituency would swiftly develop with scientists here and overseas to support the nuclear rocket program. Even here though, the Office of Space Science must prepare for the introduction of the large Re-core engine, as it would intro-duce a true golden age of space sci-ence, as I'll show later, with the intro-duction of the Super-Size Science Space Station. I know it's another dreadful name, but I want to emphasize to the reader something extraordinarily large – 20, 30, perhaps 40 times as large as the 440-ton International Space Station.

**NASA GETAWAY PROGRAM**

PHOTO CREDIT: NASA

To the dismay of many in the Unites States and overseas, NASA ended the GetAway Spe-cial and Hitchhiker programs that allowed individuals to fly payloads into space. The highly popular programs featured several 55-gallon drums, 200-pound canisters holding a variety of experiments, tucked away in unused areas of the space shuttle. Despite their demise, efforts are under way to revive the programs. The free launch program would be even more popular at home and abroad as 17,000-pounds of payload offer wider opportunity for public partici-pation than several 200-pound canisters.

However, other constituencies exist; it's unlikely that all 1.7-million pounds of payload would be dedicated to science. The pressure would be irresistible and come, I believe, from private organizations, amateur astronomers, inventors/entrepreneurs/venture capitalists, educational institutions, and the environmental or Green movement. NucRocCorp's payload/mission divisions would service the commercial sector. To respond to their demand for inclusion I see an Office of Public Missions being created out of different offices in NASA such as the Office of Education, perhaps with those in the Departments of Commerce, Education, and Energy concerned with technology transfer, entrepreneurial innovation, and education. While NASA could do this administratively, it must be created in the Democratization of Space Act so the public can provide their views and have them as part of the legislative history. Its actual composition is unimportant for my argument, but it must have equal footing with the Office of Space Science, as that is important in bureaucratic politics, but have radically different criteria for selecting payloads. Perhaps it could be as simple as: *"If you build it, we will launch it, provided it's safe."* In other words, this office would provide the public with a "free launch" of their payload without considering whether it has merit or makes a contribution to knowledge, etc.; just whether it's safe and presents no problems for other experiments or other satellites. This might seem a frivolous largesse of taxpayer money, a "free launch," but it is quite consistent with other U.S. actions to push educational initiatives and spur economic development, such as the Homestead Acts, and it unleashes different and powerful democratic forces into the space program.

## THE HOMESTEAD ACTS

Congress passed a series of laws in the nineteenth-century to spur development of federal land by those without capital. For $12, plus $6 closing costs, a settler could claim 160 acres that he would have to work for five years before gaining title. By 1932, more than a million settlers used the Homestead Acts to claim 270 million acres of federal land.

First, having a "free launch" tells professional societies and private organizations such as the Planetary and Mars Societies they could have their payloads sent to LEO if they provide it at their expense. Since they are not well endowed compared with a NASA or a Boeing, this forces them to take a fresh look at constructing payloads. I think innovations would result, of using off-the-shelf and even salvage equipment and of "thinking outside the box." Individual failures might be common early on, but I maintain these "moms and pops" outfits ultimately would get it right. Many might disagree, holding that outfitting payloads requires skills, assembly facilities, and know-how well beyond the expertise of "moms and pops," but that's just snobbery. I would not talk down the innovational skills of the "great unwashed" among the American people. If given the opportunity, I think the "moms and pops" could compete for commercial and scientific payloads, and give true meaning to the NASA mantra of "faster, better, cheaper." Many might see abuse here, or envision the Office of Public Missions launching inferior payloads doomed to fail. No, not at all. Since payloads would be privately sponsored and since the "free launch" should be only one-time, those conditions alone should weed out inferior ideas. This recognizes that people become very cautious when spending their own money, but can be quite cavalier with other people's. Moreover, I expect many firms would be created to help the "mom and pops" or private organizations with their payloads, stimulating a growth in the private sector space industry, and broadening the tax base.

Second, having a "free launch" tells back-yard astronomers and astronomy clubs if they build their own telescope, NASA would launch it. This might allow them to compete with the elite who have their own Hubble telescope and who set their own agendas on where it is pointed in the heavens, all, of course, at taxpayer expense. Perhaps space telescopes inordinately cost billions, but then the lure of having their own might prompt the amateurs to organize and innovate, with less costly, but just as capable ones the result.

Third, having a "free launch" tells inventors and entrepreneurs who have bright ideas about using Space for commercial purposes, but little wherewithal, they could reach LEO free of charge to actually test them. Just find the financing – and I noted here taxpayers might underwrite them – and build the payload. This could unleash still more creative forces in the private sector. Perhaps most would fail, but the few that succeed might become Fortune 500 companies. I suspect the NucRocCorp divisions would follow this closely and have follow-on incubator launches for promising technologies, perhaps in coordination with venture capitalists. This activity alone, I expect, will generate a lot of work for the Patent Office, and so the Democratization of Space Act should provide it with all necessary resources to handle the increase.

Fourth, having a "free launch" tells students in high schools and universities they have immediate, practical applications for the perhaps unsuccessful admonitions of their teachers to study math and science. "If you build it, we will launch it" competitions might arise in each state, developing over time a large pool of trained and motivated personnel to work in Space or other high technology areas. I also predict this would have a most ironic consequence – of wringing hands changing into slashing elbows. Now political leaders from different states would elbow each other for a better position in the launch queue for their constituents instead of wringing their collective hands in worry over a "radioactive reactor flying overhead."

Fifth, having a "free launch" tells the environmental movement they have a new tool in their arsenal to help Green the planet. I expect this to become very popular, so that dedicated Green "free launches" become the norm, and I see four main categories for them: testing of space-based ideas to mitigate global warming, to produce solar power and beam it back to Earth, to provide space-based environmental detection and monitoring, and to allow disposal of toxic materials. I will discuss this subsequently.

Finally, all these changes could occur in countries represented on NucRoc-Corp, so the forces unleashed would not be confined to the United States. Thus, the Re-core alone could begin the democratization – the liberalization - of the space program, allowing more to participate but at their own expense or with modest governmental cost. Meanwhile, the tax base would grow in the U.S. and other member countries.

**B. The Solar System Transportation Plan**

Now I will discuss a second provision the Democratization of Space Act must include, the requirement for development of a solar system transportation plan. This requires discussion of the Re-use engine and its mission scenarios, but it becomes very complex very quickly. Let me show how. I'll start with a simple assumption: The Re-core would be solely dedicated to bring LH2 and payloads for the Re-use to take elsewhere. Thus, the total would be 1.7-million pounds over its operational life. Now we're in a real quandary. A Re-use engine would need almost 12-million gallons for its 30-hour fuel life, plus payloads. That's about six Re-cores flying 100 flights each, 600 total, and that is highly unrealistic until large Re-cores appear, particularly super duty ones.

Before trying to bring order out of the chaos, I'll make it even more complicated by looking at possible missions. Flights to geosynchronous orbit, to the Moon, inner planets, outer planets, to asteroids and comets each require its own delta v (its propulsion requirements) and hence an amount of LH2. Then there's the timing of the mission, for those close to Earth (geosynchronous orbit, the Moon) timing is relatively unimportant, but for missions beyond that, it becomes increasingly important. Then the ignition sequence complicates matters even more. Instead of a constant firing to break free of Earth's gravity, perigee ignition might be selected. Here a Re-use engine would fire as it reaches the inner portion of its elliptical orbit around Earth, then shuts off as it coasts back around the Earth, only to fire again when it reaches the outer portion of the orbit. This continues until the rocket reaches its desired orbit or breaks free of Earth's gravity. The multiple restart capability of Re-use engines is ideal for perigee firing and saves LH2, but increases the time needed to carry out the mission.

Let's make more chaos. On some missions such as the outer planets, speed might be important, as scientists might not want to wait years for the payload to arrive and so opt to propel a smaller one faster, perhaps cutting the mission time in half, or more. So here we're using more LH2. Then we face the question of what to do with a Re-use engine that is near the end of its operational life. We could return it and recycle the core, as it still would be valuable, or we could get rid of it by sending it on a disposal flight. Such could be into a comet-like orbit around Earth, to return hundreds or thousands of years later when most of the radioactivity has decayed. Or it could be sent beyond the solar system or to the Sun where it would be annihilated. All disposal options could contain a payload, as it makes no sense to send

them to distant regions without gaining scientific data. These could be "free ride" missions as the engine has been amortized and must be disposed of or, if DOE wanted some money for its fuel, these could be "deep discount" ones. Then there's deadheading. This trucking industry term means to return to home base without cargo; it's costly and trucker's try to avoid it. A Re-use engine has deadheading problems; it must return to LEO, so we must think whether it could be done profitably. Some thoughts could include returning samples from a planet, moon, asteroid or comet; others could be returning malfunctioning satellites to LEO for repair or return to Earth. All of this would require more LH2. After this, we face clustering, using two or more Re-use engines to propel a much larger payload. Common with chemical rocket engines, there is no problem for doing it with nuclear, and it might be used for moving multiple container-size payloads from LEO elsewhere.

To create even more chaos, I add technological progress. A Re-use engine would not be a one-size fits-all system; different variations of it, as well as different solid core concepts, will come into existence as experience is gained, and this probably would be a swift process for trained and competent development teams. So we would have some engines optimized for geosynchronous orbit, the Moon, inner planets, outer planets, dual-purpose (electrical power and propulsion). Let me say this a different way: Forget about the other concepts on the nuclear continuum, even the humble solid core has a lot of room for growth before it reaches its technological limits. Then the large engine would appear that I discuss in the next chapter, and it could boost about 300,000-gallons of LH2 to LEO per flight. That makes our 12-million gallon requirement only 40 flights, a much more reasonable figure. It turns our "garage" into a "full service gas station." Yet even this large engine could become larger and propel even more LH2 to LEO.

We now have a big pot of chaos, stirred vigorously. Let me try to bring order out of it. I hold nuclear rockets require a systematic approach where routes and schedules are determined first, not individual missions as with single-shot chemical rockets. In other words, it requires not single-shot "mission-itis" thinking but an approach similar to an airline developing its business plan where the most popular and profitable routes are identified from the others that are less so. From that comes the best mix of airplanes to fly them, and then an infrastructure of people and facilities. This is adaptable to changing conditions, such as introducing new cost-saving technologies or later generation systems, adding or subtracting planes to fly more popular or less popular routes, and allowing more profitable ones to subsidize those less so. Yet profit is the bottom line for an airline.

Could such an approach be useful here? Could the more profitable Reuse flights subsidize those less so?

Could space operations have the flexibility to respond quickly to changing conditions, including the creation of more popular routes as well as incorporating technological advances? Could it be market-driven, with profit the bottom line? I believe the answer is yes to all these questions. How, then, could it be done?

It begins with the NucRocCorp, which would be required under the Democratization of Space Act to develop a solar system transportation plan, with its flight routes and schedules, in cooperation with NASA's Offices of Space Science and Public Missions, member foreign governments and their space agencies, and domestic and foreign private sector participants. I expect this to be a major objective to be attained during the development of first-through-third-generation engines and to be published when fourth-generation ones appear. In other words, it needs about a decade of experience to do right. To implement it, I expect NucRocCorp would divide the solar system into regions into which it establishes a transportation service and from that determines the optimum years and routes. The plan would be amended often - for example, each time a more capable engine appears (e.g., with higher specific impulses) - and it would be a key development guide, pointing out specific propulsion needs or special engine requirements.

Perhaps the first region would be the outer planets, comets or other destinations beyond the Earth's orbit around the Sun as well as the Sun itself. Such would form the basis of the schedule and missions to them would be determined not on the basis of their proximity to Earth, scientific merit or other criteria used by scientists, but on the availability of Re-use engines for disposal. These would be the "free ride" flights I mentioned earlier or at least "deep discount" ones. Scientists might fear that missions to these destinations would shrink drastically, yet the reverse would happen. Many Re-use engines might require disposal, a fact that could subsidize flights to these areas.

I expect similar thinking to be applied to destinations closer to Earth, designated the inner planets (inside the Earth's orbit around the Sun) for lack of a better term. Here the better, not best, years and routes for the missions would be determined, but not in order to dispose of an engine. Rather, I see NucRocCorp conducting marketing surveys, based on a set price per pound of payload and then after gauging the demand, dedicating Re-use engines to service this market. Initially, it might be only a single engine, but later more might be added, in effect, creating a fleet dedicated to Mars, Mercury, or Venus runs. Optimized for these runs, they would shuttle back and forth like taxicabs until they reached the end of their operational life. Then they might become "free ride" or "deep discount" flights. Moreover, sometimes this fleet would be in competition, which would drive down costs; sometimes in cooperation, such as clustering several engines to carry heavier payloads; sometimes offering different options, such as eliminating the need for electrical power for the payload by using a dual-purpose engine. Then the Re-use's speed must be considered; it could offset the less optimum years while, at the same time, offering a choice of fast, medium and slow missions with lightest, lighter, and regular size payloads. Then if some runs become popular, a division could add more engines, perhaps making the run to a planet now a melodramatic every hour on the hour. And, since talented development divisions would now exist, they likely would develop engines different from the Re-use one, perhaps nuclear ion ones, to fill in gaps in the plan. In other words, if they saw a market niche un-serviced by a fourth generation Re-use engine or by ones in the development pipeline, they could develop different nuclear systems to fill it. In sum, market forces would now determine runs to these areas with scientific merit and government funding receding. Finally, the Office of Public Missions, foreign governments, or even stockholders might underwrite launches to promote educational initiatives or competitions here, thus expanding "if you build it, we will launch it" beyond LEO. This could include the "moms and pops" that see commercial opportunity here.

The last region might be called near-Earth though what constitutes it is unclear. It might include the Lagrangian points, the Moon, or geosynchronous orbit, or it might be merely a time definition such as all missions up to a month's travel time out and back. Regardless, it would likely see the most activity, particularly after the "free launch" program takes hold, and it would have dynamic growth for science and commerce. Indeed, many flights might be venture capitalist, to test further the concepts that made the cut, to use a sports phrase, after the "free launch." Or it might be as simple as the NucRocCorp, one of its division, or an independent party issuing a prospectus to attract venture capital and then planning the mission. So determining demand in this region would be sort of a chicken-and-egg exercise, yet it is an area where costs might be defined more accurately. Undoubtedly, it won't be the $166/332 per pound I mentioned earlier, but it may be somewhere around that.

Thus, the Democratization of Space Act must require the NucRoc-Corp to develop a solar system transportation plan by the time fourth-generation small engines appear and publish it to allow citizens, companies, professional societies, scientists, and students an opportunity to see the flight schedules, the availability of space and, of course, the cost of the ride. This would be the death knell for "mission-itis" thinking, and it would be a radical departure from the chemical rocket experience, where commercial firms treat their cost figures as state secrets. If airline, truck, or train companies operated under such secrecy, they would be out of business quickly.

In summary, a striking liberalization of the space program could occur when the taboo is broken and NucRoc-Corp, a corporation-owned, government-oversight firm, is created to develop and operate nuclear engines. Its $6 billion cost would not be so fearsome to the public treasury, because industry and private citizens would supply perhaps all the funding; in other words, its financial bite would not be as big as its bark. And, if there is government funding, it must be repaid. NucRocCorp would then divide the solar system into regions and develop a transportation system and infrastructure to service it. This would democratize the space program dramatically, opening it up to more and more people, and it would become progressive by rewarding taxpayers who invest in it. All of this would broaden the tax base. And the same forces unleashed in the United States could be unleashed internationally, and since their governments, companies, private organizations, and people could participate, opposition to a "radioactive reactor flying overhead" would change into a demand to participate and benefit. This recognizes that fear always recedes as the prospect of benefits, particularly very profitable ones, increase. All this comes from the small engine. I turn next to the more astounding consequences of the 35-inch core: It allows the development of a competitive space station industry.

[i] The C5A has a payload of 261,000 pounds and a top wartime payload of 291,000 pounds, a top speed of 541 mph and a maximum operating altitude of 43,500 feet. It is more than sufficient to carry a 91,000-pound nuclear stage, plus the weight of the strap-on solid rocket boosters. That isn't a concern. Other factors are, and I discuss them subsequently.

[ii] Holmes F. Crouch, Nuclear Space Propulsion (Granada Hills, California: Astronuclear Press, 1965), chap. 12.

[iii] Let me explain. During the 1960s, Y-12 made two thousand quality control inspections on each fuel element, making their total costs $1000 per element. A Re-core's fuel will only be used for 10-15 minutes, so that will simplify the manufacturing process, making $4000 per element in today's dollars reasonable. It will not need so many quality control inspections. The rest of a nuclear engine is the pressure vessel (a small machined aluminum cylinder) a machined beryllium reflector, six long cylindrical controls drums made of boron and beryllium, six actuators (motors) to rotate the drums, some temperature sensors, one (or two) turbine/turbopump and several valves and piping and a nozzle. This is not overly complex. All could be purchased for $25 million and except for the turbopump and nozzle, most components would have long lifetimes, probably longer than 100 flights, and could be rebuilt and recertified as flight ready.

[iv] Cf. John F. Naugle, First Among Equals: The Selection of NASA Space Science Experiments (NASA SP-4215, 1991).

# Chapter 5
## The Heavy Lifter Re-Core Engine

*This chapter considers how the large engine can <u>continue</u> to broaden domestic and foreign participation in the space program and in doing so, fundamentally change its nature. It will center on creating a space station industry, which when it takes root, will cause a final transformation, a permanent shift, where henceforth a burgeoning private sector will establish and fund the activities it deems desirable while the public sector will establish their rules and regulations. That will require four new political institutions based on democratic ideals, bringing the space program into harmony with the egalitarian principles followed by many countries, and by fairness, ensuring space activities will be open to all, voluntary to all, and profitable to all.*

### Introduction

My readers may find this chapter weird and unreal, and in early drafts of it, I sensed the same thing and wondered why. It's not the technology, it's not rocket equations being true only for chemical but not for nuclear, so that cannot be the problem and the numbers say so. No, the heavy lifter I discuss here has a logic, like all technologies, and once it develops and matures it carries through to a conclusion with many consequences. It took some time, but I finally realized my problem was a mental or psychological disconnect between what I was accustomed to seeing and what the technology really offers. It is as if a brash young Rickover told the submarine admirals in 1948 his nuke sub would sail submerged around the world and under the North Pole. "Unreal, this guy is nuts" would be their reaction, because they knew their diesel/electric subs had to surface every few days to recharge the batteries and air supply. Yet the Nautilus did exactly that a decade later. It was a different technology that needed different criteria by which to judge it. I'll offer another Space Age example. In 1919, Robert Goddard published a paper about sending a rocket to the Moon loaded with magnesium, which would detonate, leaving the flash visible from Earth. At the time, his rockets had yet to claw skyward (his liquid-propelled rocket reached 7500-feet in 1935), yet he knew this goal was inherent in the technology once it developed and matured. The numbers said so. However, relying the conventional wisdom of the era, the <u>New York Times</u> ridiculed the idea; to its credit, the paper apologized in 1969 when men landed on the Moon. Goddard, unfortunately, died in 1945. Today, we evaluate nuclear rockets with chemical-rocket criteria when we must not; we must eliminate round-peg-in-square-hole thinking. That is what the Nautilus experience should teach us. Goddard's example should teach us to stop thinking small, of single shot missions, that flawed "mission-itis" approach, when we must think big and bold, like the lunar landings, and know something along the lines of the following discussion is inherent in heavy lifters. The only question is when and how it appears.

With those thoughts in mind, here I

will discuss the heavy lifter and its economics, then analyze its technical infrastructure and part of its political one. I began this discussion in the last chapter and will finish it in chapters 7 and 8. Finally, apart from this, I will consider the most radical of all heavy lifters, the single-stage-to-orbit, as it is a logical conclusion of heavy lifter development. It will radically change the space program.

## I. The Heavy Lifter

For my argument I assume a Re-core heavy lift engine would have a 35-inch core, a size standard for most of Rover/NERVA. But, it is not the only choice and I refer the reader to the boxed text for additional information. It would have 1000-seconds of specific impulse, weigh 15,000-pounds, be 30-feet long, and have an 180,000-gallon LH2 tank

---

### Heavy Lifter Engine Specifications

Many heavy lifters are possible as the B-4 core design is inherently flexible; it can be scaled up to develop bigger reactors, from 20-inch core diameters to perhaps a limit of 55-inches. Most cores during Rover/NERVA were 35-inches, but Phoebus 2A was 55-inches. Also, the B-4 can increase in power density (the thrust obtained from a given core volume). Now engine specifications will depend on many factors. One factor is the boost: How to get it to 100,000-feet so it can fire to LEO. For my argument, I assume a 35-inch core pushing a stage weighing 395,000-pounds and a cargo plane developed to carry it. That is within today's technology, but the drop may be a problem. A stage so heavy may make the plane uncontrollable when the separation occurs. Yet it does not appear any fundamental laws of flight prevent it, so it might be just a difficult task that no one ever saw the need for, and a properly structured research and development program could solve it in a straightforward way.

Another factor is the market and competition among engine divisions. Different size heavy lifters are probably inevitable, the same as there are 18-wheeler trucks for long hauls and smaller ones for around-town deliveries. Some divisions might specialize here, so smaller stages might appear weighing only 200,000-pounds, perhaps making a cargo plane drop easier. This might mean an engine in the 100,000-pound thrust range (2000 MW), with a 30-inch core. Then a sea launch is possible, using solid rocket boosters fired from a ship to take the stage to 100,000-feet. This may mean an engine in the 200,000-pound thrust range (4000 MW), and core around 40-inches. Finally, launch from a U.S. controlled island in the Pacific is possible, again using solid rocket boosters. Some divisions might see this as highly profitable for sending ultra heavy payloads to LEO. This may mean a Phoebus 2A engine in the 250,000-pound thrust range (5000 MW). With increases in specific impulse and power densi-

ties, these could become, in principle, heavy lifting behemoths. I'll say this differently: A Saturn V with a first-generation NERVA I (1500 MW, 825-seconds) as its third stage could boost almost 500,000-pounds to LEO; these would be fourth-generation systems and go from there, so they could have gargantuan payloads, thus changing the cost per pound numbers. While this sounds wonderful, they would have many problems, starting with the stage that lifts them to 100,000-feet and with their bulky LH2 tank. However, the use of slush hydrogen, discussed in Appendix B, could mitigate this problem by allowing smaller tank sizes.

Still another factor is heat and the propellant. As engines reach over 3000° C, propellants other than LH2, such as ammonia, methane and water, may become viable. Above that temperature, these propellants, which have weights around 18, disassociate into atoms; maybe it's not as good as hydrogen, but for nuclear rocket first stage applications it might be just fine and double the lifting power of a solid rocket first stage. In other words, if a solid first stage booster has a specific impulse of 330-seconds; these could be 600 or 700-seconds, and not only be more powerful, but also allow the dimensions of the stage to shrink. So Krafft Ehricke's ammonia first stage nuclear rocket, boosting a LH2 second stage to 30-miles up, might just be prescient. As both engines would return in their cocoon, two stage nuclear rockets could boost bulk payloads, like LH2, to LEO quite inexpensively. Or the new space station industry might find itself thinking in modular form, with each module weighing perhaps 500 MT or more, with several being assembled together in LEO. Extraordinarily large, yet low cost, space stations could result. Thus, starting small engine development first gives time for these issues to be studied and for the market to develop. When it does, it would decide what is developed.

(the turbopump running at 200-gallons/second for a 15-minute flight). The tank and LH2 would weigh 120,000-pounds (90,000-pounds for LH2 and 30,000-pounds for the tank) while the payload 150,000-pounds, for a total stage weight of 295,000-pounds, plus the solid rocket boosters. As the flight would last about 10-minutes, the tank weight and LH2 amounts are too generous, but I am not trying to design a heavy lifter stage - only introduce a new way of thinking about them. Like the smaller Re-core, this large one would require a cocoon at 10,000-pounds to bring it back to Earth, making the effective payload 140,000-pounds. It should have a 100-flight rating, and so could boost 14-million pounds to LEO or about 6400 MT. Design could begin when NucRocCorp was formed, but the approval to develop should be given only after fourth-generation small engines appear, i.e., after a decade to give time for infrastructure to develop.

---

## Specifications of Fourth-Generation Heavy Lifter Re-core Engine

**Engine**
150,000-pounds of thrust (3000 MW)
15,000-pounds weight
1000-seconds of specific impulse
200-gallons/second of LH2

**Stage**

| | |
|---|---|
| LH2 tank | 30,000-pounds |
| Payload | 150,000-pounds |
| Cocoon | 10,000-pounds |
| LH2 | 180,000-gallons |
| 90,000-pounds | |

Totals:

| | |
|---|---|
| Engine | 15,000-pounds |
| LH2 tank | 30,000-pounds |
| LH2 | 90,000-pounds |
| Cocoon | 10,000-pounds |
| Payload | 150,000-pounds |
| | |
| 295,000-pounds | |

---

Much of the infrastructure needed to support this large Re-core need not be discussed since that developed for the small engine should be straightforward to adapt to the larger. Strong and competent development teams, with a decade-plus experience, would have solved problems with the smaller Re-core/Re-use engines and now could confidently address these. However, three problem areas appear to stand out. First, the cocoon might be difficult since the engine will be 30-feet long; this might require jettisoning the nozzle in Space, leaving only the cocoon to hold the engine, which would be about the size of a 55-gallon barrel. Another problem might be disposing of its LH2 tanks. Their size might prevent their return to Earth, and their growing number might pose a clutter and safety problem in LEO. The best approach is to view them as assets to be used and explore this in detail before concluding they are debris to be eliminated. Finally, developing a cargo plane would be a key unknown, particularly one able to carry and drop a 295,000-pound stage, plus perhaps another 100,000-pounds for the strap-on solid rocket boosters, and one with a very large LH2 tank. That's 395,000-pounds, and it's beyond the capacity of the C-5A, but not beyond today's technology. Indeed, the giant Russian cargo plane, the Antonov 124, or the newer Airbus might carry it. Still, it's a real problem, but I think it's more a case of "No one has ever asked us to do it," than "No, it can't be done."

Developing this engine should cost much less than the original NERVA, whose cost in 1971 dollars was to be $1.1 billion or about $4.4 billion in today's dollars, because the B-4 core is scalable and since strong and competent development teams would exist. So costs should drop significantly and likely go through only one-generation system before the engine outlined above was ready and this might take only two to three years. Perhaps those costs might be only $2 billion, plus another billion for new test facilities. All that would be needed would be suitably sized turbines, turbopumps, and nozzles, and much of this knowledge would exist from the original NERVA program as well as a decade's experience with the smaller engines. The other components such as the

pressure vessel, control drums, and actuators would require development but that should be neither long nor expensive. They are not big-ticket items. I'll say $50 million for these. So a large Re-core might cost only $2 billion, plus another billion for test facilities, with each operational engine $50 million. Each engine would contain 1870 loaded fuel elements and at $4000 per element, that's $7.5 million per core or $750 million for 100 flights. The total so far is $3.8 billion.

Now I turn to the launch infrastructure and its costs. Though it's uncertain now, I'll assume a giant launch plane can be built for $500 million to carry this heavy lifter stage and its very large LH2 tank to mach 0.7 and around 50,000-feet. I assume each launch would cost $4.3 million ($500,000 for the aircraft, $1.5 million for the LH2 (at $8 per gallon x 180,000-gallons), $1.3 million for the tank, and $1 million for the solid boosters. The total for 100 flights then is $430 million.

So now our costs are: $2 billion for engine development; $1 billion for test facilities; $50 million for a Re-core and $750 million for 100 new cores; $500 million for a cargo plane; and $430 million for 100 flights. The total is $4.73 billion and for that amount 14-million pounds could be taken to LEO. That's $338 per pound if all development costs are amortized. Counting just the engine, its cores, and launch costs, the figure is $1.23 billion to launch 14-million pounds. That's $87 per pound.

This $87 figure is stunning, but I repeat my warning of the last chapter. It's theoretical. Its value lies in its implications. It implies the small-large engine progression is best, as development cost and time may be quite small; indeed, the cost lies predominantly in fuel fabrication and test facility construction and much less in pressure vessels, turbopumps, control drums, and nozzles. Also, the low cost implies public tax dollars need not be involved at all, as NucRocCorp and its divisions can obtain the money on their own or from taxpayer investors. This is not a large amount for such stunning consequences. Without a doubt, I expect market pull to be quite dynamic by now, with different engine development divisions competing furiously, as all would have completed their market surveys. I do not doubt that different sized heavy lifters would appear, so the $87 per pound figure might just be the mean, with lighter duty ones somewhat higher and super duty ones somewhat lower. Perhaps most important though, this $87 figure implies something else, something that overrides any fear of accidents, something so powerful that it commands the attention of the public and political leaders: The political, economic, social, and other consequences of heavy lifters. They justify the program because now a fundamentally different and democratic world order could be created, one in preparation since the formation of NucRocCorp and the "free launches," and one bringing the prospects of world peace and a Green planet. I laid some groundwork to discuss Leo Szilard's world peace idea previously; I'll discuss it a bit more here, and finish it in chapters 7 and 8. I'll finish my discussion of the Green theme in the next chapter.

## II. Introduction to the Space Charter Authority

I've mentioned a space program conducted with chemical rockets gives primacy to the executive branch, while one conducted with nuclear rockets gives the legislative branch primacy, and it will oversee the private sector conducting activities with its own money. This transformation would have started with the Democratization of Space Act, and when NucRocCorp begins development of small engines, its boards of directors must begin debate and dialogue over the political institutions to govern heavy lifters. NucRocCorp's decision to approve heavy lifter development would signify to all the time for debate has ended, and the time to create institutions has begun, thus completing the transition to a peaceful and democratic space program. Of the four political institutions needed, I will discuss only the purpose and technical aspects of the Space Charter Authority here.

### A. The Space Charter Authority's Primary Objective: Colonization

I start by asking two fundamental questions: What does the Space Charter Authority do and how would it do it? In other words, what's its purpose? The answer to the first it is quite simple: It determines how heavy lifters are used in the colonization of Space, and its jurisdiction extends to the end of the solar system. Membership in the Authority then gives each country a voice in shaping the nature and pace of colonization. However, it would not have any jurisdiction over any country conducting a space program with chemical rockets. They could establish a lunar base or Martian outpost if they desired, though it would be economically ruinous.

Well now, this is quite a claim, and at first glance, for any individual, authority or group of institutions to claim dominion over the solar system seems preposterous, insane. That would be true if premised on chemical propulsion yet a closer analysis reveals something very different. My assertion is sound, as it is based on the nuclear continuum and rocket equations, but all technologies on it are not realizable yet. Only the solid core is. So only time is involved, time for these other technologies to come into being and mature. To deny this is to assume technologies, such as fusion, are permanently denied to science.

Recognizing this time factor, as one of its first acts, the Authority should establish zones of expansion, linking each to the standards necessary for the most economic development possible. For example, a single criterion of specific impulse might be used. Here Zone 1 might be limited to LEO (less than 300 miles) and 1000-seconds of specific impulse; Zone 2 to geosynchronous orbit and 1500-seconds;

Zone 3 to the Moon and 2000-seconds; Zone 4 to the inner planets and 3000-seconds; and Zone 5 to the outer planets and over 3000-seconds. In other words, as progress is made along the nuclear continuum, so progress is made in opening up more zones for activity. Though it might seem absurd to think of a base on Pluto now, perhaps the furthermost zone, the appearance of fusion engines with specific impulses perhaps in the millions would alter that view dramatically.

This may be too simplistic as other criteria could be used, e.g., economic in the sense of reducing the per pound cost of launching payloads into Space and returning them to Earth, say from the $87 to $50 per pound. So when that is reached, a new zone is opened. Or it might be a mixture of management, labor, and capital requirements to ensure the safe and efficient development of new zones. I have no favorite and assume as others begin thinking about this many views will appear, prompting a vigorous and insightful debate that leads to the development of precise zones and specific criteria. For my argument though, I assume Zone 1 is limited to less than 300-miles and 1000-seconds of specific impulse. This eliminates a large Re-use engine and many missions popular with space advocates such as lunar outposts and manned Mars since they would be in zones not yet opened for development. Still, I see little conflict as zones are based on private, not public money, and if advocates could present a viable, privately funded effort for them, the Authority should consider their approval. In the end though, the specific criteria are of secondary importance. What zones do, however defined, is bring discipline, focus, order, and a worldwide and collective pool of capital (discussed shortly) to bear on a single area of development at a time. For my argument, the basic objective in Zone 1 then would be to fill LEO with space stations, and to build on Earth the infrastructure to nurture and sustain them.

Finally, the zones must be harmonized with the solar system transportation plan for small engines and the transition point would be the "garage," which would be transformed into a "full service gas station." It should be big and robust enough to house service personnel permanently, creating the first privately sponsored astronaut core. It should be able to handle all types of nuclear engines, with an ample supply of spare parts on hand, and hold perhaps a million or more gallons of LH2. Some engine and operations divisions might specialize in servicing such facilities and quite likely, super duty heavy lifters would arise to service this market. It could be quite profitable.

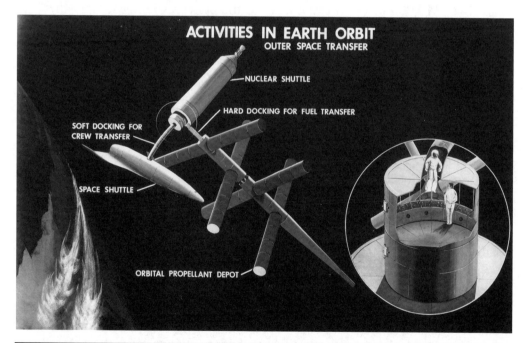

**GAS STATION**

PHOTO CREDIT: NASA

Small Re-use engines would require a "garage" in LEO to provide vital services, but heavy lifters allow it to be transformed into a "full service gas station," with large supplies of LH2 to fill up nuclear engines returning from elsewhere, various bays to service different size engines, and mating/de-mating capabilities for payloads. It would likely have a permanent crew with expertise not only for nuclear engines but perhaps also for payloads. Repairing malfunctioning satellites or upgrading them could be a lucrative business. Perhaps it might even become a satellite production facility, with an extensive inventory of parts to configure a payload to the customer's specifications. This could reduce space insurance costs markedly, as the payload and engine to take it somewhere would already be in LEO. This would reduce risk, too. In the 1960s, NASA contemplated a precursor to this full service gas station for NERVA.

## B. The Authority's Means of Colonizing

Now my second question is: How would the Authority go about colonizing the solar system? I begin by noting heavy lifters change supply and demand. If it costs a 200-pound man $30 million to go to LEO via chemical propulsion, the demand is extremely low. Thus, the space program could putter along with a mix of mostly government with some private funding, as it has done since 1957. At $87 per pound, the demand curve radically changes as it would cost a 200-pound man about $18,000 to go to LEO, and as more capable heavy lifters are introduced, this figure would go lower. This means the far-fetched dream of real space tourism, not a roller-coaster ride, would not be so far-fetched, and the dreams of those who seek to use Space for purposes other than science, communications, or Earth resources would not be so far-fetched either.

Thus, I expect a dynamic increase in demand, and the Authority must establish the legal vehicle for allowing it. Also, this demand would be capital intensive; that must come from the private sector, leading investors to expect profits. Governments neither could sustain such development with their annual appropriations cycles nor manage it effectively for profit – in fact, they should not engage in profit-making ventures. Yet I maintain capitalism and nationalism alone should not determine the course of this new activity. In other words, just because the United States and a few advanced countries could build nuclear heavy lifters doesn't mean only they should use them. They must be used to achieve other goals and objectives, such as world peace and Greening our planet.

I see the charter as the means of achieving it. In essence, a charter grants rights or privileges to a group or organization to carry out certain activities. Those who know world history after 1500 are familiar with the charter as the principal means to acquire the capital necessary to colonize the new worlds discovered by Vasco da Gama, Christopher Columbus, and other legendary sea captains. A few include the Virginia Company, the Massachusetts Bay Company, the Danish, Dutch, French, and English East India companies and the Hudson Bay Company. Here *individual* sovereign governments established the charter, but private capital underwrote it, and while most companies were highly successful and profitable a great danger lay within this approach. The companies had vast grants of power such as negotiating treaties, waging war, administering justice, and issuing currency. That led to egregious abuses, and imperialism and wars resulted. Perhaps Senator Anderson had this in mind when he said in 1956 the colonization of Space should be international.

There are different types of charters. Normally, a *single* government creates a private legal entity for a purpose within its borders, including buying and selling property, offering a service, marketing a product, entering into contracts, suing and being sued, hiring and firing, borrowing and lending money, and so forth. Today this is called a corporation, and it has been highly successful since the onset of the Machine/Industrial Age, with competition among corporations a benefit, and regulations such as anti-trust laws to prohibit their unpalatable consequences. However, allowing that model (a single government creating a corporation) to proceed into Space via heavy lifters will cause more problems than it solves. It would exacerbate North-South tensions – where a single or several advanced countries proceed with little reference to the less advanced – and impede the spread of democratic values.

A single government, though, can create a public corporation to carry out functions within its borders deemed important for the citizenry. It works, as the Tennessee Valley Authority (TVA) shows. A variation of this works in the Space Age with Comsat/Intelsat, which 120 governments created. However, I am not confident a public corporation would work here. TVA, for example, is an independent agency with an annual congressional appropriation, yet the goal must be to decrease and eliminate the need for public money while creating an extensive and privately funded space industry. Also, Comsat/Intelsat's benefits do not filter down to the people whose tax money funded their government's involvement except if they own stock. Finally, I do not believe a single government can advance democratic ideals and fairness - many must unite to do so.

I'm getting ahead of myself here, as I will complete this a political discussion in chapters 7 and 8.

So I return to the charter, and the Authority must establish a clear set of articles to govern a chartered entity. Some would be strictly of a business or commercial nature while others would be unique to space operations. I'll focus on the latter to begin a dialogue about them. The most important articles would pertain to safety, particularly since flights would be manned, and here the charter must require a standard language for all launch and space operations and a standard measurement scale. This would leave the chartered free to use their own language for business and commercial activities, and it recalls the unmanned NASA satellite that failed when it used components made to two different measurement scales. Also, the articles must ban inordinately dangerous activities though defining such may be difficult, as just being in LEO is dangerous. Still, some commercial operations might constitute such an undue risk that a charter must not be granted. Another key article would be the legal code and judicial process under which disputes and issues would be resolved, no matter whether they are between the Authority and the chartered, between the chartered and other governments or private parties, or between the chartered themselves. The law of the sea might prove useful here, the oceans and the ships sailing on it being somewhat analogous to manned operations in Space. The experience of the International Space Station where diplomacy resolves disputes is not relevant here – too slow and cumbersome and its impartiality is suspect. Perhaps existing legal systems could be adapted to this new situation, or perhaps courts and other impartial review processes must be established to adjudicate disputes. I tend to think the latter. Other key articles must allow the chartered to conduct normal business practices such as buying, selling, leasing, bartering, trading, merging, mortgaging and even going out of business, provided they would not endanger safety or encumber the Authority with liabilities. Perhaps a bond as a condition of a charter might insure against irresponsible actions. The charter's length must be considered – on Earth it's in perpetuity or until the corporation goes out of business or is taken over or merged, but I don't think this is a good idea for space stations, as new managers or owners might arise that skimp on safety - and so 25-years might be a useful period as most business ventures recoup their investment in that time. Linked to this is charter renewal, as the chartered certainly would want assurance they could continue a profitable enterprise for another period of time, and charter reviews, as the Authority certainly would want to review all charters and make adjustments as necessary. Five-year reviews might be a useful, but the regulatory regime I discuss in chapter 7 might eliminate the need for the review and renewal process entirely.

### III. Space Charter Authority's Technical Infrastructure

Here I will discuss the technical infrastructure that would be required for the Authority to carry out its purpose. The reason is quite simple. Just because a heavy lifter Re-core and a Space Charter Authority and its paper charters would exist does not mean there would be any rush into Space, unlike what would occur with the smaller Re-cores. In fact, probably nothing would happen. Why? Because only a void would exist in Space, not an infrastructure. That must be created in the same way national governments did in years past to build the infrastructure for the railroad, automobile, telephone and airplane. I'll leave aside for the moment *who* should do it and *where* the money should come from to focus on the technical infrastructure itself. This would involve two space stations (though I dislike that phrase), a passenger vehicle or bus, and a nuclear power reactor. The first two must not be configured like the International Space Station, which weighs 440 MT, can support a permanent crew of six, floats in Space with zero gravity, has solar panels for power and whose central purpose is to be an orbiting laboratory to conduct scientific and commercial experiments. Though hardly economical and marginally productive given its $100 billion cost, the ISS is the best that can be achieved with chemical propulsion. Heavy lifters, however, require different thinking that must be based on creating a competitive space station industry as quickly as possible, consistent with safety. Here the phrase "space station" might be defined more precisely according to its purpose.

**THE ISS**

PHOTO CREDIT: NASA

The International Space Station, an on-going cooperative project of many governments, will cost an estimated $100 billion to complete. Its money derives from each government's annual appropriation cycle, and it operates under an international treaty with layers of other, more specific legal understandings. Both cause numerous delays that increase costs. Yet the ISS is the best that can be achieved with chemical propulsion. Under the charter system, the Authority would create and authorize a legal entity to conduct designated activities, but its financing, construction, and operation would come from the private sector, which understands that delays cost money and profit. Moreover, charter employees would be results oriented, as they know the private sector rewards performance and punishes the lack thereof.

## A. The "Demonstrating and Proving" or "Put Up or Shut Up" Station

The first of the two space stations might appear as early as fifteen years after creation of NucRocCorp, and take the most critical part of the research and development process to LEO. Called by many terms, I will call it the "demonstrating and proving" phase. This pivotal step is well known to most industries: Agriculture, airlines, automobiles, pharmaceuticals, and military ordnance. Here a new technology, product, or process proves its merit in full field-testing, not laboratory experiments; here problems with a full-sized prototype, not a mockup, are uncovered; here the transition from theory to practice, from promise to reality occurs or does not. The ISS cannot do this; it cannot house prototype-manufacturing processes. It is simply too small and ill suited for any meaningful activity other than keeping the hope of Space alive. Any promising applications it uncovers would still require "demonstrating and proving" before beginning full-scale commercial applications. Likewise, its small crew is ill suited for the task since most are astronauts or scientists. Engineers are needed here, particularly industrial ones, and many more than the six on the ISS. The most famous example of this need, of this transition, is the few scientists who took atomic energy from paper theory to the Chicago pile (the first reactor) in 1942; yet it required the U.S. Army Corp of Engineers, the DuPont Corporation, and thousands in industry to take it to concrete reality a couple of years later.

**THE VON BRAUN SPACE STATION**

CREDIT: NASA

In 1952, Wernher von Braun developed a space station concept made of nylon that would be inflated in orbit and rotate to provide artificial gravity. It has inspired other inflatable concepts that would be quite dangerous for humans because of cosmic radiation and solar flares. They are attractive nonetheless because of the limited ability of chemical rockets to propel tons to LEO economically. The photo above shows a 24-foot prototype inflatable space station developed in 1961 by NASA's Langley Research Center and the Goodyear Aircraft Corporation. Heavy lifter nuclear engines could take thousands of tons to LEO economically, allowing a variety of safe, rugged, and inexpensive designs to appear, the start of a new industry.

Since a single heavy lifter might take 14-million pounds to LEO over its operational life, perhaps half that amount, 7-million pounds or about 3200 MT, should be the weight of a new type of space station, one to "demonstrate and prove" other space station concepts. As noted, the goal is to create a competitive space station industry. Scientists would exist here only as part of an engineering team. This station should be thick-skinned and robust, since weight is not a factor, and provide accommodations for many - perhaps fifty might be a useful figure to begin this discussion. It should rotate to provide artificial gravity, perhaps the same as on Earth to minimize health problems, and it should provide shelter for protection from cosmic radiation and solar storms. It should have ample power, perhaps 50 MW from a small reactor – the second development project. Most important, it should have a large amount of easily configurable space for working in gravity-free, partial, or full Earth gravity environments. Linked to that should be large, well-provisioned machine and maintenance shops and medical facilities.

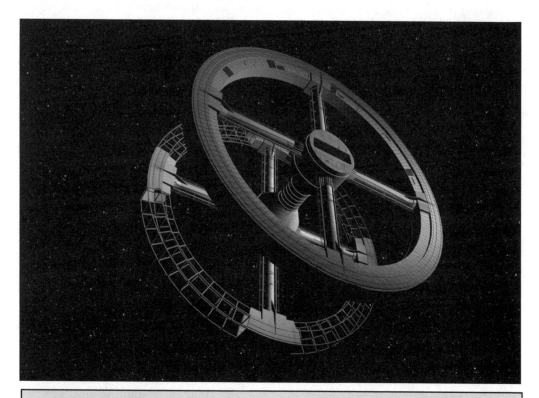

**2001 SPACE STATION**

CREDIT: Apogee Books

Wernher Von Braun designed a similar space station to the one shown in the movie *2001: A Space Odyssey*, an extraordinarily large 900-feet in diameter. It would weigh thousands of tons, well beyond the capability of chemical rockets to lift economically, but heavy lifters make such large structures technically and economically feasible.

## The Nuclear Power Plant

The reactor should be small, compact, and rugged. Perhaps 50 MW is a good figure with which to begin this discussion. It should be able to be increased in power, though just adding more reactors might satisfy greater demand. Its safe operation and ease of maintenance are a given. Russia has developed space nuclear reactors while the U.S. worked on them early on, but then emphasized the use of radioisotopes for space satellite applications (which could not supply 50 MW for this type of space station).

I expect NucRocCorp's division to fund and develop these reactors; the cost is unknown now, but the divisions would have well trained teams developing the small engines, so "tweaking" them for other earthly markets is quite likely. I've already mentioned adapting their technology for electrical power and process heat applications, but the 50 MW systems have emergency (disaster/blackout/ brownout) or remote power applications. These could be very profitable. This would be a $21^{st}$ century version of the USS Sturgis, a WWII Liberty ship modified to contain a small nuclear reactor, which provided power to the Panama Canal for almost a decade in the 1960s and 1970s. Under this, DOE and the Federal Emergency Management Agency might establish a program whereby these power plants are kept on DOE sites, but flown to an area having an emergency or dire need and linked to the grid. Once the situation stabilizes, they would be returned. Since they would be small, I expect they could be shielded without undue weight and carried by cargo planes and trucks to the point of use. They also could be of use in remote areas, such as providing power to the scientific outposts in Antarctica.

Solar cells must be rejected. They are best for unmanned, low demand and near Earth applications. Moreover, their wide surface areas leave them vulnerable to strikes from all sizes of meteorites and space debris, a sharp contrast to the ruggedness, compactness and high power output of a reactor. The ISS has solar cells, but floats in Space, and that poses fewer structural problems compared to space stations that will rotate to provide artificial gravity. In essence, solar cells are best suited for space programs conducted with chemical rockets.

I predict when Congress passes the Democratization of Space Act, with the small-large engine progression, formation of NucRocCorp, and creation of a NASA Office of Space Stations, private sector interest in space stations would intensify everywhere and keep pace with engine development, first participating in the design of the "garage" for the Re-use engines, then the "full service gas station," then progressing to this one, a big one, a tough one, a safe one, with ample power but not devoted to science, not designated as a transfer point for other space ventures (though that would happen with other stations), and not dreamily dedicated to be a "space colony," whatever that means. Uncharitably, it might be called a "put up or shut up" station and not a "demonstrating and proving" one. That is why I gave two names for it, with this one having a nasty edge to it.

**The Office of Space Station Design**

When Congress passes the Democratization of Space Act, it must include an Office of Space Station Design whose purpose would be to create a space station design, construction, manufacturing, operation, and repair and maintenance industry. Its first projects should be the "garage" and "full service gas station," followed by the "put up or shut up" and "convincing stations." It should form partnerships with NucRocCorp, other regime member space organizations and with domestic and overseas private sector firms to build these, so it might have public and private funding. However, once these have been built and a space station industry established here and abroad, this office must revert to conducting pioneering research and development and safety work. The private sector henceforth would build the space stations it desires. One exception, however, might be the Super Size Space Science Station (discussed subsequently). The office must develop and build test facilities if such are needed and this might be similar to the giant wind tunnels NASA operates to study aerodynamics. One such facility might be a giant 200,000-ton press to form large segments of the space station out of single blocks of metal to eliminate the need for welding and other joining processes, which weaken a structure, and to simplify their construction in LEO. Super duty heavy lifters might carry them into LEO.

Why? Very simple: The literature abounds with concepts of space tourism, space hotels, space colonies, space manufacturing of new products in a gravity-free environment, space power generation back to Earth, space this and space that. Most want public money to bring them to reality, which is one reason why they remain concepts. In addition, the visionaries, inventors, and entrepreneurs who benefited from the "free launch" might find their work added to this list of promising concepts. Perhaps all could be profitable, but I doubt it. All have been stymied by the extremely high cost of reaching LEO, but after a decade's experience with the smaller engines, the heavy lifter's economics would be much better defined, giving more reliable figures on the costs of taking large payloads to LEO. But that, too, is not enough; people must be taken to and from LEO safely and so now I come to the third development project: A bus, man-rating it into something similar to NASA's Crew Exploration Vehicle/Orion that's being designed to carry six astronauts, but clearly much larger. Or more likely, it might be similar to Lockheed Martin's Crew Exploration Vehicle that proposed the use of a lifting body shape to return to Earth. So the ability to carry fifty people might be a useful design objective for a first-generation system. If each passenger weighed 200-pounds, the total is 10,000-pounds, leaving the remaining 130,000-pounds of payload for their provisions for an extended stay in LEO, and for the structure of the bus itself. It would separate from the Re-core engine, dock at the space station, and be used for regular or emergency returns to Earth, perhaps to an airport if it uses the lifting body. The heavy lifter would dock at the "full service gas station" before returning separately to Earth. I expect NucRocCorp's divisions would develop the bus, and its development might start during the small engine era, making "minis" to haul perhaps a dozen up to the ISS whose mission, in turn, would change to doing experiments or other tasks for the two "garages" and "put up or shut up" station.

This bus would be used over and over, with later-generation systems probably following quickly. So now key data points would exist: The heavy lifting Re-core engine and its cargo plane, the bus, the "put up or shut up" station, and the nuclear power reactor, from which to evaluate the profitability of these concepts. Many would be eliminated right away or postponed until the economics became better. For those that remain, the other key part would involve demonstrating and proving they could work; after all, that is the purpose of the "put up or shut up" station, that's the purpose of being nasty, to force those who talk into action. This station then should be Authority-owned and *a fee-for-service facility*. Concepts would be scaled up to near commercial size and tested, giving another key data point from which to determine whether to proceed to a full-scale commercial enterprise.

Let me pause now to summarize the transformation that would be occurring. With charters, the business and legal aspects of a space venture would be specified. With the heavy lifting Re-core, the economics of moving large quantities of materials into LEO would be determined, and I expect several NucRocCorp divisions would compete for the business, probably by operating different size heavy lifters. With a bus, the economics of moving large numbers of people to and from LEO would be determined, and I expect several NucRocCorp divisions would compete here as well. With an Authority-owned "put up or shut up" station and its nuclear power reactor, a facility would exist to test concepts near commercial size and to derive data on the profitability of a full-scale business. Since this station would fee-for-service, its introduction nears the end of a step-by-step transition from public to private financing of space activities, a process begun with the Democratization of Space Act. But, unfortunately, I believe it still would not be enough. There must be proof the "put up or shut up" station works by a full-scale example of something developed from it and operated profitably. In other words, the private sector must see hard evidence everything could work as conceived and give its financial backers a certain hope of profit.

## B. The "Convincing Station"

How would or could this next step work? Many options would exist such as computer chip manufacturing or pharmaceutical production, but such, I sense, would badly miss the mark of what would be needed: Publicity, a space station designed to advertise the attractiveness, potential, and profitability of LEO ventures, a station to allow Earth-Space thinking to take firm root and grow. I do not have a good term to describe it, but the inarticulate phrase "convincing station" might come close. Let me explain why.

As its first project, the "put up or shut up" station would prove those technologies required to construct and operate a "convincing station" that houses several hundred for several months at a time. It would appear several decades after creation of NucRoc-Corp. Thus, it should be large, perhaps 1.4-million pounds or about 6400 MT in weight, double the size of the "put up or shut up" station, with ample radiation protection for these many inhabitants. Maintaining their health must be a key objective here because people of various ages and health conditions will occupy the station. Here I eliminate the lengthy and costly medical review and training process developed for the elite

corps of astronauts and scientists who fly in Space today. Instead, a more democratic phrase would apply, more akin to "If you can walk, you can fly," though that would need tempering, but perhaps not much. Senator John Glenn was in his late 70s when he flew to the ISS. With less fit and less trained of all ages flying, this new station would require a proven medical capability, able to treat and cure most of the maladies several hundred people might encounter over several months, including heart attacks and strokes as well as toothaches and the common cold. Medical science has been part of Space since Sputnik, but this would require it to be a hospital in LEO with surgery, recovery, and therapy facilities and a permanently trained staff. Similarly, the technologies for providing food, air, water, and sanitation for so many people would need proving and demonstrating. This should be as closed a loop as possible, with the nuclear power reactor used to recycle and re-circulate as much of these vital commodities as possible. The ISS's open loop system, requiring a constant re-supply of life-sustaining commodities to a few occupants, is unattractive, uneconomical, and dangerous for sustaining hundreds for months at a time.

With this "convincing station" I have reached the point in my argument where the democratization of Space really begins. Its two hundred occupants would not be the elite who have dominated Space since Sputnik, but private citizens who fly as part of flights organized around different themes, all of which would be dedicated to facilitate the transition to private funding. The first flight might be for business awareness, with two hundred executives from the manufacturing, pharmaceutical, hotel/tourist, or other industries. This would force them to integrate two dimensional Earth-Space thinking into their business models, or if they choose not to, to face the consequences of their competition doing so. This could be followed by dedicated flights of executives from the financial, insurance, and venture capital sectors, then flights by those in the burgeoning space station design industry, including its labor workforce. Linked to this would be flights to train people in creating and operating a Space Regulatory Commission.

(In transitioning from a publicly to privately funded space program, a regulatory body is critical to ensure safety, and its imprimatur indispensable in obtaining the support necessary to create and sustain a business venture in LEO. It's a hands-on, daily function that must be created and become part of any space venture from beginning to end. Thus, future space regulators must gain experience in the construction and operation of the "put up or shut up" and "convincing stations" and from this develop their regulatory regime).

Other flights could be dedicated to other parts of the business cycle, but who flies and in what order does not matter. What would be critical is that there is a logical progression to develop a base of capital, management, and labor from which an organization could develop its business model, make a decision to construct a commercial station, and have a pool of knowledgeable workers from which to draw to build and operate it. All this is premised on the conservative nature of the human psyche, of its unwillingness and hesitancy to commit in a major way to something new unless convinced otherwise. It is true on a personal level where we have engagements before marriage, or test-drive a car, or inspect a house before purchasing. It is true on the business level where executives pore over vast amounts of data before committing to build a new plant or start a new venture. Only the foolish buy a pig in a poke; everyone else seeks to be convinced.

Others, however, might need convincing to see the importance of the two-dimensional thinking, and perhaps the third year of the "convincing station's" operation could be dedicated to them. After all, a democratic space program must allow all in society to participate. Perhaps the first group might be several hundred from the news media, first to publicize the potential of LEO, but more importantly to create a base of informed opinion makers to critique LEO development. The fourth estate has a legitimate role in a democracy, and as Space is democratized so should their ability to analyze it. Writers might come next to experience Space viscerally and a rebirth of literature might result. After all, why should mankind's greatest classics be confined to some guys who knocked around the Mediterranean in a wooden boat, another who fought a mythical dragon, or one who harpooned a white whale? Spending time there could sharpen their insight and allow their imaginations to soar on what man is doing and can do in Space. I would include artists and composers here also. And looking down on the bright Earth and into the black heavens might create a new perspective among Protestant, Roman Catholic, and Orthodox theologians, and perhaps lead to the end of the schism plaguing Christianity. The same might be said of the division between Sunnis and Shiites in Islam, and the religions of the Near and Far East might benefit from a new perspective.

Finally, the public at large must be allowed to fly, and here I differ with many who see the public as eager to do so. No, I believe the literature overestimates the public's support, much of it blinded by science fiction books and movies or photographs from the ISS or Hubble telescope. I hold it more accurate to view the public's attitude as similar to its initial doubts and fears of the automobile and airplane a century ago. And here, even with buses flying dozens at a time, it is one thing for the public to support them, since they go for commercial gain or professional reasons and know the risks, and quite another to fly themselves or allow their sons and daughters to do so, particularly for a holiday. In society as a whole, the more venturesome will fly first and

as this happens safely, the public's attitude will change. Perhaps within a decade it would be similar to today's acceptance of the automobile and airplane.

I see two ways to allow the public to fly. The first is to sell tickets and make a profit. I suspect this would be popular with affluent adventurers in the industrialized countries but not so in the less affluent ones. So a lottery might be used: The Authority would sponsor it and split the revenue among the Authority member governments or payoff the bond it issued to buy the station (below). A variation might be regional lotteries. Here member governments from a geographical area such as Africa or South America would conduct it and split the revenue. If this became popular, it would show that the private sector space tourism will be a profitable business, after which the Authority should change the mission or divest itself of both the "put up or shut up" and "convincing stations" and not compete with the private sector. Sell both, payoff the bond if it hasn't already been paid, and split whatever is left among the member governments and citizen stockholders. That would symbolize the complete transition of the space program from public funding to private.

This might take several decades, perhaps disappointing those hoping for immediate gains, but I hold it is realistic and follows the examples of the automobile, airplane, telephone, and railroad, which all needed similar lengths of time for their infrastructures to develop and become integrated in the larger economy. Moreover, I feel charters will be quite profitable but do not know if their earnings might equal that of the Dutch East India Company, which never paid less than a 12 percent dividend for nearly a century. I do know capitalism rewards its risk-takers generously and those who invest reap the reward. I have described a way to democratize it to allow even the most humble to participate.

Now I will discuss who will own, operate and fund the development of these four infrastructures. As noted, the Space Charter Authority must own them for a time, but some other entity must operate them. But who? This is difficult to discuss, because the small engines and "free launches" would have generated great interest in Space as would anticipation for the heavy lifters and space stations. In other words, many established as well as startup firms would want in on the action. This is one way it might work. When the Democratization of Space Act passes, NASA would have an Office of Space Stations. I expect the space agencies of other member governments to be restructured likewise to support a space station industry, with the first fruits of this labor being the "garage" and "full service gas station." So a partnership would develop between NucRocCorp, space agencies, and a burgeoning private sector space station industry. Out of this mix would come an entity to design, construct and operate the "put up or shut up" and "convincing stations." In truth, I expect NucRocCorp's divisions to be primary here, restructuring to enter this market or entering into partnerships with other private sector firms. The Authority should grant this entity a long-term lease to operate the two space station, but structure it in such a way to develop a knowledgeable work-force to operate the space stations to follow. So it's sort of a learning tool also.

Now, who would fund it? I expect NucRocCorp's divisions and its tax-payer investors, perhaps with venture capitalists and a developing space station industry, to fund the "put up or shut" and "convincing" stations. This would establish their market positions via patents, intellectual property and know-how. In other words, government money would not be needed except as a last resort and as it pertains to funding the NASA Office of Space Stations, whose purpose would be to sustain this industry over the long term. I also expect these same divisions and taxpayer investors to eagerly fund the bus, because it is intrinsically linked to the heavy lifter, and would establish their market positions. I would expect the same for the nuclear power plant, as developing it might require no more than "tweaking" from the systems developed for Earth applications. All four would be sold to the Space Charter Authority, which as I noted, would issue a bond to pay for it and recoup it via "fees" for issuing charters initially, then with tourism, lottery proceeds, or other revenue producing streams. When the four have fulfilled their purpose, the Authority could dispose of them as outlined, and payoff the bond, if required. In sum, I believe all this would occur with private money. In other words, I think by the time heavy lifter development begins, market pull would be extensive and dynamic, and eliminate the need for government funding, other than as a last resort. But if it is, it must be repaid.

### ***The Single-Stage-to-Orbit***

As progress continues along the nuclear continuum, with ever more powerful engines developed, the single-stage-to-orbit (SSTO) vehicle becomes feasible. This holy grail of spaceships takes off from the ground, flies into Space, then returns to refuel and fly again and again and again. Everything is airliner economic and neat and tidy - no booster rockets or empty fuel tanks being discarded to clutter up LEO, causing a safety problem. Just a landing strip anywhere will do, thank-you-very-much, or so the theory goes. A chemically propelled SSTO is not possible, as the recently canceled, billion-dollar boon-doggle Venture Star proves. Another gold-plated nail, this vehicle was to use LH2/LOX and weigh 2.63-million pounds. Yet if it could work, it could carry only 56,000-pounds to LEO –

about 2.13 percent of its gross takeoff weight, still a tiny payload fraction. The Venture Star followed on the heels of the National Aerospace Plane proposed by President Reagan in 1986 to reach LEO or Tokyo in two hours. That gold-plated nail cost taxpayers $1.7 billion before it was cancelled in 1992.

In contrast, Robert W. Bussard developed the first realistic SSTO concept in 1961, the ASPEN, a plane that would take off horizontally with turbojet/ramjet engines; upon reaching 100,000-feet its nuclear engines would fire boosting it into LEO. Then it would return to base and fly again. He revised ASPEN in 1971, as he realized reducing the core's diameter caused a sharp reduction in shielding requirements.[i] This is another way of saying increasing power density and specific impulse could decrease a reactor's core dimensions, and this in turn could reduce the size and weight of the shielding needed for the crew and payload. Here he also introduced the concept of the disposable core – the Recore idea I use here – to mitigate the problems in servicing the plane when it returns. Just remove the old core and the plane becomes like a jetliner that crews can fly or service normally. To return to LEO, just insert a fresh core. He had various ASPENs that would weigh 500,000-pounds and carry from 30,000 to 80,000-pounds to LEO, depending on their configuration. In other words, payload fractions up to 16 percent.

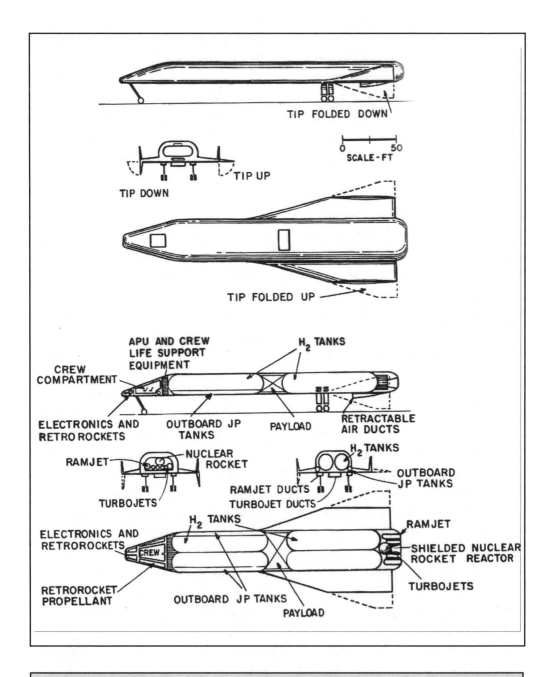

**ASPEN**

CREDIT: LOS ALAMOS NATIONAL LABORATORY

Robert W. Bussard conceived of ASPEN and updated it in 1971, a turbojet/ramjet SSTO that could carry up to 16 percent of its 500,000-pound gross takeoff weight to orbit.

To my knowledge, no one has conducted nuclear SSTO studies since 1971. Fourth-generation heavy lift engines might power an SSTO, and since turbojet and ramjet engines have advanced in power, and many lighter weight materials are available for aerospace uses than at that time, it is conceivable an SSTO might be lighter than 500,000-pounds yet carry more payload than in Bussard's studies. On the other hand, a nuclear SSTO might require more powerful engines at 1100- or 1200-seconds, with power densities around 4000 MW in a 35-inch core. Or solid cores might never be powerful enough because of the radiation protection requirements for personnel and payload.

To illustrate this, I'll compare heavy lifter-buses with SSTOs. Upon reaching LEO, the bus would be detached from the LH2 tank/stage and then the engine from the tank/stage. It might remain in its cocoon and be housed in the "full service gas station" before its return to Earth. Sometime later, the people would return to Earth separately via the bus. These operations would mitigate but not eliminate the need for radiation protection. In contrast, an SSTO might house its personnel and payload for longer in Space and have a piloted return with them to Earth. Here, starting about 100,000-feet, radiation would be air-scattered back into the SSTO and on landing, it would require moveable shielding to allow personnel to disembark and robots to remove the engines.

Obviously, for this flight sequence,

radiation protection is more important than for buses. There would be exposures, but amount is the question, and that involves determining the all-critical design standard. One of zero exposure might make a solid core SSTO too heavy and unable to fly, but to even think that is absurd since Space is intensely radioactive and people flying there would be exposed. In Space, one cannot make radiation exposures zero just as one cannot make exposures zero on Earth. It's impossible. Another option might be to adopt the nuclear industry standard of 5-rem per year, but this might still result in an SSTO too heavy to fly. Still another option might be to use the total dose a person might receive from a working career in Space and from that deduce a level appropriate for each SSTO exposure. This would be a variation of the National Council on Radiation Protection's recommendation of a career whole body dose of 100- and 150-rem for 25-year old female and male astronauts, respectively.[ii] This approach might lead to appropriate criteria for designing and operating SSTOs.

However, radiation protection is more complex than just adding weighty shielding, and may involve clever design to place personnel and payload as far from the engines as possible, and the development of a crew/payload compartment that would separate from the SSTO when it reaches LEO. Then it would fly separately to the space station. So here we're using time to minimize exposure, and the maximum would be 10 to 15-minutes. This leaves the SSTO in orbit for a month or so, during which time the reactor's radioactivity decays to more manageable levels. It might involve the use of lightweight radiation-absorbing materials to block or deflect as much as possible before it reaches the personnel and payload, and the adoption of expedited flight operations to make the transition from 100,000-feet to the ground and the disembarkation as swift as possible, thereby limiting the exposure time. Or the core might be removed in the "full service gas station," placed in a cocoon and sent to Earth, with ASPEN returning then with its passengers to any airport. Also, today's designers would have room for innovations since each heavy lift, 3000 MW engine would weigh 15,000-pounds, not the 40,000-pounders Bussard used in his studies. And as SSTOs are studied, different flight profiles would be developed that limit exposures. Still, as noted, even more powerful solid cores might be required or a solid core SSTO might be impossible. Only engines with specific impulses starting at 1500-seconds might make it feasible. Nothing exists from which to make a more informed statement. No one has studied this since 1971.

So when Congress creates NucRocCorp, it must mandate SSTO studies, with a cash award to spur competition and creative thought. Why? Somewhere along the nuclear continuum SSTOs become feasible, with stunning consequences. These bigger hammers, these more powerful engines, can overcome any design restrictions or shielding weight penalties yet carry an ever-larger payload fraction. That changes the economics of reaching LEO. Also, SSTOs would eliminate the clutter problem of jettisoning large LH2 tanks in LEO. And all nuclear engines (except for some fusion concepts) would emit radiation, so that problem would not disappear. It is just overcome. These are technical factors. Most important would be SSTOs dazzling political and social consequences; they now justify its development. Wow! That's a stunning assertion. How?

Today, many would say the deepest fear would be an SSTO crashing and scattering radioactivity to the environment, a nuclear Challenger accident, raising strident opposition by the public and other governments. Perhaps this is the reason why nuclear SSTOs have been ignored. But I see, instead, the deepest concern to be premature enthusiasm and strong competition by foreign governments and their peoples for them. This seems ironically absurd, but it is not. The SSTO's most important need would be for launch and emergency landing facilities as isolated as possible to mitigate harm to the public from contingencies during take-offs and landings; yet they would need long launch and landing corridors. The use of launch and landing corridors over the continental United States would be unsuitable for early generation SSTOs; the Challenger accident scattering debris over Texas and Louisiana would remain a vivid reminder. So new facilities would be needed, but where? Islands in the Pacific would get attention, as they are quite isolated and if located in proximity to the equator allow the SSTO to take advantage of the Earth's rotation to improve the payload fraction carried to LEO. Their isolation, however, would make them unattractive, as their logistical costs would be high; but most important, their political voices would be weak. Others would clamor to host the launch and landing facility, and would drown them out.

Of those voices, I think the loudest would be from the governments of Columbia, Ecuador, and Peru, in the northwest portion of South America. Each has isolated, sparsely populated mountainous areas or plateaus several miles high near the equator, yet logistically they are central in the Western Hemisphere. Takeoffs from there would take advantage of the equatorial boost, but the flight path itself could occur over water or the Amazon jungle basin with ramjet/turbojets, with the nuclear engines firing as the SSTO flies over the Atlantic Ocean. Then the flight path would continue over the sparsely inhabited countries of equatorial Africa, the dense jungles on the western side and upland plateaus on the east. I suspect many countries would vie to host an emergency-landing site and perhaps more than one might be desirable. After Africa, the flight path would be nearly all over the Indian and Pacific Oceans, with emergency-landing sites on different islands as appropriate. And the wide Pacific would allow a long glide path for the return to base in Columbia, Ecuador, or Peru. Essentially, this would create a launch and landing corridor within a band 10° north and south latitude of the equator, most of which is over sparsely inhabited regions or over water.

**WORLD MAP**

The northwest part of South America has unique advantages for an SSTO spaceport. The sparsely inhabited regions are about two miles high and close to the equator, both of which have advantages for the launch. However, the area's principal advantage lies in the launch and landing corridor that can be created, about 10 degrees in latitude on either side of the equator. So the flight path to LEO would be over sparsely populated areas or over water while the return from LEO would be over the wide Pacific Ocean, whose many islands could serve as emergency landing strips.

I have called them launch sites but "spaceport" more accurately describes the full range of activities that would occur at these democratic gateways to Space. Why? The Re-cores would eventually disappear, made obsolete as progress is made along the nuclear continuum. As this continues, even more capable SSTOs would be introduced, with vastly decreased fuel requirements to achieve orbit. Ultimately, an SSTO would need only a fraction of its total weight as fuel, perhaps just several percent, and a stark contrast to the 80-90 percent of chemical rockets. Initially, a first-generation SSTO might have a payload fraction of only 16 percent, but later-generation ones would see ever higher payload fractions, to 20, 25,or 50 percent for gas cores (if feasible) and even higher for fusion. This sliding scale would alter economics profoundly. As costs drop, activity and opportunity multiply. In time, the Space Charter Authority would open up more zones for development. This would increase demands on the "spaceport," so what starts out as an isolated launch site for safety reasons would have many industries surrounding it to support space activities, something similar to what surrounds major airports today. This would demand a skilled workforce who in turn would demand the amenities of life such as decent housing, schools, entertainment, and opportunities for professional advancement. Ultimately, a "spaceport" would have

an educated middle class that previously saw only a few suffering a hard-scrabble existence. And as demand increases, so would the need to create other "spaceports," perhaps in the upland regions of Africa or elsewhere in South America. This would continue for at least a century.

Thus, I view premature political enthusiasm for a century-plus of growth and prosperity in hitherto undeveloped areas as the most serious concern, as it might bring inadequate planning and hurried construction, leaving a poorly thought out "spaceport" inside its gates and a chaotic conglomeration of facilities outside. To avoid this, I hold Congress must require NucRoc-Corp to develop standards for the first SSTOs, such as they will not be developed unless they can take a minimum of 15 or 20 percent of their gross take-off weight to orbit, have reliability equal to the airline industry, and expose their personnel and passengers to radiation levels deemed acceptable and reasonable. These studies should also include all aspects of "spaceports" and emergency facilities in prospective host countries. Both should have long gestation periods to think through the problems while gaining experience with small engines and heavy lifters. At some future time then, if feasible with solid cores, SSTO development could start; then "spaceport" development could begin with confidence.

[i] Robert W. Bussard, "ASPEN-An Aerospace Plane with Nuclear Engines," LA-2590, Los Alamos National Laboratory, September 1961; Bussard, ASPEN, April 1971 (author's files).

[ii] "Guidance on Radiation Received in Space Activities," National Council on Radiation Protection and Measurements, NCRP Report No. 98, July 31, 1989.

# Chapter 6
## Nuclear Disarmament, Nonproliferation, Nuclear Energy, and the Environment

*This chapter discusses how NASA-operated Re-core and Re-use engines could create new, effectively measurable nuclear disarmament initiatives that strengthen the nonproliferation regime. Doing so, however, would require new legislation that leads to a treaty creating a new but small international agency. As these initiatives took hold, they would have important energy and environmental consequences, which lead to a Greener planet. Finally, a separate bilateral initiative with Russia could add to this nonproliferation regime and introduce democratic values deep within Russia. Doing so, however, would require US and Russian legislators to adopt a new perspective on their traditional roles (appropriating money, levying taxes, approving treaties, overseeing their executive branches) by creating a special legislative district that contains a new, joint for-profit corporation over which both legislatures maintain oversight.*

## Introduction

In the past two chapters, I discussed the growing privatization of the space program that restructures NASA away from being an operational agency toward supporting research and development for commercial space endeavors, particularly space stations. However, the public sector still requires its own launch program, but now focused on quite different objectives: Nuclear disarmament, nonproliferation, and environmental ones. This would be the first expansion of two-dimensional thinking, and work on it should begin as soon as the Democratization of Space Act is passed. In discussing it, I must review nuclear matters since 1975, and then analyze its current and future state of affairs to place my argument in its proper context. Then I will consider the legislation needed to redress the situation. After that I will discuss new thinking for the environmental movement and a U.S.-Russian nuclear initiative that has broad-reaching implications.

## I. Review of Nuclear Affairs

I start by summarizing the nuclear fuel cycle. It has seven distinct steps, and begins with mining to extract uranium from the ground, continues with its conversion to a form suitable for enrichment, then its enrichment in U-235 content to 2-4 percent for most reactors and 93 percent for weapons and other applications, such as nuclear rockets.[i] The fourth sees its fabrication into fuel elements, the fifth sees their use in a reactor, the sixth sees their reprocessing to recover the remaining uranium and newly made plutonium from the irradiated fuel, both of which have economic value and can be reused in a reactor. This is called a mixed oxide fuel cycle (MOX), the term signifying the plutonium and uranium are mixed together and fabricated into fuel. Reprocessing also sees the recovery of irradiated wastes that fall into two broad categories, high-level and transuranic wastes, which must be handled. This is the seventh and last step, disposing of these wastes. In general, high-level wastes are thermally hot, tend to be neutron and gamma ray emitters and thus are hazardous, but collectively they have short half-lives, meaning their intensity decreases with time. Here three hundred years is a useful, collective figure. Transuranic wastes are easier to handle, as they emit low power alpha and beta particles, but collectively, they have long half-lives, into the thousands of years. Since WWII, the U.S. and other western nations with nuclear programs stored both in underground tanks (the Soviet Union was more cavalier here); now they are encapsulated in an impervious material such as glass for long-term storage above or below ground.

Here I focus on reprocessing, the sixth step, and two types of plants exist, one for weapons and one for power programs. Their biggest difference lies in the front-end of the plant. Why? Reactors producing plutonium for weapons have their uranium enclosed in aluminum and operate at much shorter duty cycles, i.e., the fuel doesn't stay in the reactor long. This is called low burn-up fuel, and produces weapons-grade plutonium. In contrast, power reactors have their fuel enclosed in a metal such as zirconium or zircaloy that is impervious to radiation damage; it stays in the reactor a year or more. This is called high burn-up fuel, and produces reactor-grade plutonium. The terms weapons- and reactor-grade plutonium are artificial, as the latter can be made into a nuclear explosive, but that has uncertainties. See the boxed text for additional information.

## Weapons- and Reactor-Grade Plutonium

The irradiation of uranium in a reactor produces many isotopes of plutonium. Of these, plutonium-239 is the most useful for nuclear weapons while the others have characteristics that make their explosive use difficult and unpredictable. However, the isotopic composition of the plutonium varies with the length of time it is exposed to neutrons. So weapons-grade plutonium is formed by keeping the uranium fuel in a reactor for a short period of time, such as several months. Typically then, such fuel would contain the following isotopes:

| | |
|---|---|
| Pu-239 | 93.5 percent |
| Pu-240 | 6.0 percent |
| Pu-241 | 0.5 percent |

Keeping the uranium in a reactor for a much longer time, such as a year or more, forms reactor-grade plutonium. Typically then, such fuel would contain the following isotopes:

| | |
|---|---|
| Pu-238 | 1.5 percent |
| Pu-239 | 58 percent |
| Pu-240 | 24 percent |
| Pu-241 | 11.5 percent |
| Pu-242 | 5 percent |

The Pu-240 is highly undesirable as it emits more neutrons than Pu-239, and this causes concerns that a nuclear weapon using it would pre-initiate. In other words, it would be a nuclear fizzle, not a bang, but it still could be a big and deadly fizzle. It would give a nuclear yield though its size would be unpredictable. The militaries of nuclear-weapons states want to know precisely the force of their weapons, and hence prefer the use of weapons-grade plutonium. For a rogue state or terrorist group, however, that precision may not be a concern.

Now what does this have to do with reprocessing? Everything: Weapons-grade aluminum clad fuel dissolves easily in acids while reactor-grade fuel clad in zirconium/zircaloy does not. It is impervious to acid. So a reprocessing plant for weapons-grade fuels dumps them into acid-filled vats, where all dissolve into a liquid, and then proceeds to separate the plutonium, uranium, and high-level and transuranic wastes. However, a reprocessing plant for reactor-grade fuels has a shear, a guillotine-type blade at the front end to chop the long fuel element rod into several inch long chunks. They are dumped into vats of acid, and the process begins to separate the plutonium, uranium, and wastes, with the metal chunks removed separately. So one plant has a shear and the other does not. (Other differences exist between the two, but that discussion is not pertinent here).

Building reprocessing plants is called closing the fuel cycle and was standard for the nuclear weapons states from 1945 onwards and later for the nuclear power industry. A split occurred, however, after India detonated a nuclear device in 1974. The United States abandoned plans to reprocess irradiated fuels from its power reactors; this decision assumed the recovered plutonium was a strategic, weapons material and therefore should not be recovered. The premise for this action was moral: If the United States foreswore separating more plutonium, others would follow the U.S. lead and refrain from initiating weapons programs. Known as the once-through fuel cycle, here plutonium would be left in the irradiated fuel and then it would be buried somewhere for thousands of years.

In the following years, however, many countries (Iraq, Iran, Libya, North Korea, and Pakistan) began nuclear weapons programs without reference to this moral example, and the United States has searched for a way to dispose of its growing inventory of irradiated fuel, trying unsuccessfully to open a permanent site under Nevada's Yucca Mountain, next to Jackass Flats. Moreover, many countries including Britain, France, Japan, and Russia disagreed with this moral argument and saw plutonium as an energy commodity and indispensable step in waste management. They constructed immense, multibillion-dollar commercial plants to reprocess their own fuels and those of other countries and recycle it into power reactors - the MOX fuel cycle. This has gone on for decades, leaving the United States isolated while the plutonium problem is growing exponentially.

The Japanese commercial reprocessing plant is located in Rokkasho in northern Japan (left) and the British THORP plant is in Sellafield, England, near the border with Scotland and next to the Irish Sea (right). THORP operated for about a decade when it had an accident several years ago classified as a category 3 on a scale of 1 to 7. It was shut down, but will reopen once repairs are completed. The French nuclear complex in Cap La Hague has two reprocessing plants, UP2 and UP3, which have been operational for years. The British and French have operated a fleet of dedicated double-hulled tankers to carry irradiated fuel from the Far East to their reprocessing centers for decades.

I must explain. Since 1945, the United States and Soviet Union produced about 100 MT of plutonium each for their weapons programs, and the other nuclear weapon states a much lesser amount, about 50 MT. The United States ceased military plutonium production in the 1980s while Russia continues to produce it as a by-product of their need for electricity; the reason is technical and not because of weapons requirements. The Soviets developed a power reactor type that uses aluminum-clad fuel, but it does not store well and corrodes readily, causing radiation contamination concerns. So it is reprocessed. In the 1980s, the Soviets planned to build their future nuclear power program around the light water reactor, but after the Soviet Union's dissolution, the Russians have been unable financially to implement it. The plan included a large, uncompleted reprocessing plant called RT-2 in Krasnoyarsk in central Russia to handle light water reactor zirconium/zircaloy fuels. Meanwhile, nuclear power programs in countries such as England, France, Germany, and Japan matured, and today there are over 400 reactors in operation worldwide. They make plutonium daily and their total already vastly exceeds the total produced for nuclear weapons. That plutonium, however, is in spent fuel form, but much of it is being reprocessed and recovered by reprocessing plants in Britain, France, Japan, and Russia. That exponential growth and recovery is unstoppable and will continue for at least the next 50-75 years, making this the Plutonium Century. Inventories will be measured in thousands of tons in recovered and spent fuel form.

Governments have established different means of dealing with this growth. The United States cooperates with Russia to dispose of much of its weapons-grade plutonium, including its recycle in the old-style Russian reactors. The United States will do likewise with much of its weapons-grade plutonium, recycling in its nuclear power reactors as MOX, while it searches for a permanent repository for irradiated fuel from its once-through fuel cycle. But the prospects of Yucca Mountain opening are remote. Overseas, MOX is planned for England, France, and Japan, and other countries for reactor-grade plutonium, but under even the most optimistic scenarios, supply will greatly exceed demand. Other schemes for handling recovered plutonium include denaturing, i.e., adding a material that makes it unsuitable for weapons or explosive purposes. Unfortunately, this can be reversed and is not a permanent solution to any unauthorized, malevolent use. A variation of denaturing is to mix the plutonium with high-level and transuranic wastes and bury

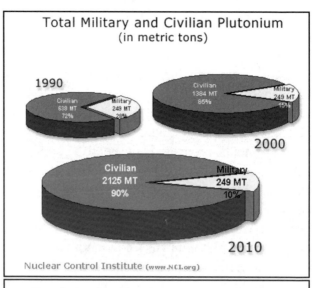

**Total Military and Civilian Plutonium**
(in metric tons)

1990 — Civilian 638 MT 72%, Military 249 MT 28%

2000 — Civilian 1384 MT 85%, Military 249 MT 15%

2010 — Civilian 2125 MT 90%, Military 249 MT 10%

Nuclear Control Institute (www.NCI.org)

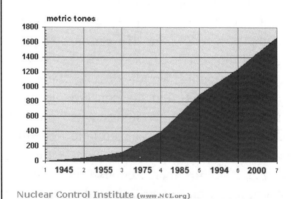

**Total Worldwide Plutonium Production**
(1945-2000)

metric tones

| | 1945 | 2 | 1955 | 3 | 1975 | 4 | 1985 | 5 | 1994 | 6 | 2000 | 7 |

Nuclear Control Institute (www.NCI.org)

**PLUTONIUM GROWTH**

CREDIT: NUCLEAR CONTROL INSTITUTE

Plutonium growth will be dramatic over the next 50-75 years and cannot be halted, as many countries rely on nuclear power for their energy requirements. The small wedge on the pie chart shows the United States and Soviet Union produced about 100MT each for their weapons program and the other nuclear weapons states about 50MT combined. Everything else will come from nuclear power reactors, as the growth curve chart indicates. Though it ends in the year 2000, the curve is still dramatically up, almost a straight 45 degree line up, as more than 400 reactors worldwide produce plutonium daily.

it somewhere, but this doesn't really solve the problem as both the mixture and plutonium can be recovered. Thus, a situation exists where plutonium inventories are growing exponentially but where different countries have intractable differences on it, the United States seeing it as a strategic material that must not be recovered but sent to a underground repository for long-term storage. Most other industrialized countries see reprocessing as an indispensable waste-management step, and the recovered plutonium as a commodity that secures their energy independence. Adding to the gravity of the situation are the rise of terrorism in the world, and the concern that 25-pounds of reactor-grade plutonium could be diverted and made into an explosive with the power of the Hiroshima bomb; and the increased use of fossil fuels in burgeoning worldwide economic development, and the concern that this has created a greenhouse gas effect that could cause catastrophic global warming. The last point highlights the need for increased use of nuclear power, which contributes nothing to the greenhouse-gas effect.

## II. The Plutonium Disposition Act

I hold the nuclear rocket solves this multifaceted problem as it allows two-dimensional thinking – Earth and Space – to emerge and render obsolete all one-dimensional, Earth oriented thinking. How? Very simple: Dispose of plutonium in Space. This idea has been around since the beginning of the Space Age, but is impossible to achieve or prohibitively expensive with chemical rockets. As this new thinking emerges, it has the important policy consequence of ending the loggerheads with which the United States finds itself in the world and allows it to take the lead to establish a new plutonium disposition regime. To begin this debate, I will present my vision of how it can happen.

This new thinking must be premised on the nuclear continuum as it allows an ever-decreasing cost of moving large payloads into Space. However, only the solid core is viable now. This limits options, the most desirable of which would be to take the fuel directly from a reactor and launch it into Space (more precisely from a storage pond next to the reactor that holds the rods for several years, allowing its radiation to decrease to lower levels). That would solve everything, plutonium and wastes at once. However, since the fuel would be radioactive, it must be launched in costly spent fuel casks that can weigh over 100-tons. This appears beyond the capability of the solid core and even if less costly casks were designed for super duty solid cores, it probably leaves this option highly uneconomical as only a few irradiated fuel elements, out of thousands produced every year, might be launched at a time. Hence, the time for this option is not yet.

This leaves the option of launching plutonium recovered from reprocessing, and this might seem highly attractive. For example, a heavy lifter could carry 1.4-million pounds to LEO over its 100-flight lifetime; that's 6400 MT of plutonium. That's more than the projection, so the problem seems solvable with perhaps only 25 flights. Unfortunately, that is impossible, but to explain why I must consider one unique trait, among many, of element 94 on the periodic table: I must discuss the term <u>bare sphere critical mass</u> or bare sphere crit. Most other elemental metals can exist in ton quantities without problem – iron in automobile engines and bridges, lead in sailboat keels, and so forth. This is impossible with plutonium no matter whether it is reactor- or weapons-grade, because both emit neutrons constantly. This means a self-sustaining chain reaction can occur if enough plutonium is present in the right configuration under the right conditions. This is what the term bare sphere crit means. For example, 11-kg or 25-pounds of metallic weapons-grade plutonium in the shape of a round ball will go critical; in other words, a self-sustaining chain reaction will occur, but not a nuclear explosion. For metallic reactor-grade plutonium, a bare sphere crit is 13- kg or 29-pounds. In plutonium oxide form, it is between 35-90-kg (77-200-pounds). In reprocessing, where plutonium is in liquid form, bare sphere crits are avoided, inter alia, by using holding tanks with critically safe geometries, that is, tanks shaped to avoid criticality. So a tank might contain hundreds of kilograms of plutonium in liquid form, but it likely would be thin, several inches thick yet quite long and wide, making it more like a slab than a tank.

Thus, disposing of plutonium in Space would require first and foremost avoiding bare sphere crits during launch and mission operations, and accident scenarios. To consider this, I will use the small engine's 17,000-pound payload, not the heavy lifter's 140,000-pounds. Here a payload container must be developed to withstand the stresses and strains expected during launch and mission operations, including accident scenarios. Then choices would be made on the matrix material to hold the plutonium; here it actually aids the process. It likes to form compounds, so many matrix materials would be available from which to select the best for a launch into Space. It might be something like a molten glass-plutonium slurry that is just poured into the container and allowed to harden. Neither seems to be a difficult development task. So now let's say at the end of this development, a payload might hold 1000-pounds of plutonium out of the 17,000-pounds available, the rest being the matrix material and weight of the container. Still, that would be highly significant, 100,000-pounds of plutonium or 45MT over 100 flights of a single Re-core engine. But this must not be viewed as a one-time affair. There would be many flights over the years and decades, with the flights synchronized with reprocessing to keep plutonium inventories on Earth at an agreed level.

Technically, when Congress passes the Democratization of Space Act, it must direct DOE, NASA, and the Nuclear Regulatory Commission to study the disposal of plutonium in Space. The overall design parameter should be the payload expected from fourth-generation Re-core and Re-use engines, e.g., 17,000-pounds and a cargo plane launch, and the work should include all the steps required to flight-certify the container. Quite likely, that would require a research and development and certification effort using the first-through-third generation small engines. The disposition profile would require early attention as many options exist: Storing the containers in Earth orbit, as an energy bank; landing them on the Moon; also perhaps as an energy bank; sending them into a comet-like orbit, or out of the solar system, to Alpha Centauri, the star nearest Earth. Obviously, most scenarios would require other nuclear engines with various propulsion capabilities, and I expect the studies would produce many options on the most cost effective, the most attractive, and those with the most and least delta v (or propulsion) requirements. However, science might have a say here, as it makes no sense to send plutonium to a distant region and not receive any scientific data back. So this might be a plutonium version of the "free ride" discussed earlier. I don't wish to prejudge that review, but I expect sending plutonium to the Sun would emerge as the best option as our giant fusion reactor will annihilate the plutonium completely. Nothing will remain. But this might be the most costly from a propulsion perspective, yet cost might not be as important as showing the world the plutonium is gone forever.

While technical work continues, as part of the Democratization of Space Act, Congress must establish an Assistant Secretary for Space Nuclear and Nonproliferation in the Department of State, a slot reserved for a Foreign Service officer. His or her office must prepare its views on the new plutonium disposition regime and would have a critical role in implementing it.

To this regime I now turn my attention. It could be subsumed under the Outer Space and Non-Proliferation treaties, but both are ambiguous. For example, Article IV of the Outer Space treaty declares "States Parties to the Treaty undertake not to place in orbit around the Earth any objects carrying nuclear weapons or any other kinds of weapons of mass destruction, install such weapons on celestial bodies, or station such weapons in outer space in any other manner." The plutonium certainly would not be in a weapons configured form but in a matrix material and payload container that prevents criticality. Moreover, the Treaty would not apply if the plutonium goes to an energy bank. Some may allege, however, that even in a matrix form, the plutonium is a weapon of mass destruction and therefore banned. Still, this regime might be included under Article I; it states the exploration and use of Outer Space shall be carried out for the benefit and interests of all countries. Disposing of excess plutonium and the temptations it might cause and the risks it might incur certainly would be in the interests of all countries.

Likewise, Article VI of the NPT

requires its parties "to pursue negotiations in good faith on effective measures relating to cessation of the nuclear arms race at an early date...." Obviously, if countries disposed of plutonium in Space where it could not be retrieved or retrieved easily, or where its annihilation would be certain, it would be an effective measure to limit the amount available for weapons or for unauthorized diversions. However, some countries might object, arguing Article IV permits the peaceful development of atomic energy and that the plutonium they possess is only reactor-grade and dedicated for energy uses, not weapons. In brief, these treaties could be made to include a new regime, but I feel they have too much ambiguity and would not be the right path to follow. Moreover, both reflect the conditions of the 1960s. Indeed, Article V of the NPT permits the use of peaceful nuclear explosives but those disappeared in the 1970s and their reappearance is unlikely, and the only rockets that existed then were chemically propelled, so the Outer Space treaty reflects just its capabilities. Nuclear rockets, I submit, require a totally different treaty that reflects its capabilities. To apply the former to the latter is another instance of round-peg in square-hole thinking.

Thus, I believe new legislation would be required, the Plutonium Disposition Act, to mandate changes in the U.S. nuclear program as well as lead to the new international regime. Clearly, the most important would be to end all work on the once-through fuel cycle.

In other words, kill the Yucca Mountain repository. I expect many in Nevada would greet that with joy, but I also expect it would be criticized vociferously. Yet it would be absolutely necessary. The United States could not lead the creation of a new regime while, in effect, refusing to participate in it once created. That's what adherence to the once-through fuel cycle means. The position would be untenable: U.S. plutonium must be part of the new regime. This means reprocessing, closing the fuel cycle.

Four options exist for doing this. The first is to allow the private sector to build the facilities, but that would generate a complete lack of response. Private industry invested hundreds of millions in building commercial reprocessing plants in the 1960s only to have the government refuse to license them and adopt the once-through fuel cycle instead. The old adage of "once burned, twice shy" applies here. Moreover the Morris, Illinois, and Barnwell, South Carolina, plants have been cannibalized or dedicated for other purposes, so the prospect of retrofitting them seems nonexistent.

The second is to upgrade some or all of the existing facilities at DOE's Savannah River and Hanford sites and reprocess there. If this is feasible (it may not be, as these plants were built for low burn-up fuel), it is attractive: The cost could be lower than building new facilities, a base of reprocessing and waste-management expertise still exists, strict security measures are in force, operations might start within a decade, the utilities would have sites in which to ship their irradiated fuel, and the government would avoid lawsuits as it has charged utilities a fee for the once-through fuel cycle repository but failed to open it. That has forced utilities to spend millions to store the fuel on-site. In fact, the United States would benefit by creating a for-profit reprocessing industry, thus relieving itself totally of its obligation under the once-through one while broadening the tax base. Henceforth, U.S. utilities would have to contract for reprocessing services.

How? Since DOE sites are GOCO (government-owned, contractor-operated), one way is to allow the current contractor to retrofit the existing plants at its expense, and begin commercial reprocessing operations. However, since contractors face periodic contract renewal, this may be disadvantageous if DOE did not renew the contract or if the firm decided to pursue other commercial endeavors. A better way is to carve out the reprocessing plants and request bids from the private sector for a dollar-a-year, fifty-year contracts for rights to them. Here any residual money left over from the utility tax would be dedicated to refurbishing the plants, yet whoever won the bid would incur all other costs to upgrade them to handle power reactor fuels. Its reward would be a guarantee it could recoup its investment and make a profit over the next fifty years. Let me state this a different way: Nuclear power plants will have 30-, 40-, or 50-year operating

lifetimes in which time they will produce sizeable quantities of irradiated fuel that must be reprocessed. That guarantees a market, it's fixed, and most likely a profitable one. Furthermore, if Congress allowed taxpayer participation similar to that in NucRoc-Corp, this could be even more attractive since here would be an interest-free source of money. Yet taxpayers could share in the profits, and this would be equitable since their tax dollars built the original reprocessing plants in the first place.

The third option is to allow utilities to sign reprocessing contracts with England, France, and Japan. I will discuss Russia, the fourth option, shortly. This should happen anyhow as it introduces competition into the reprocessing market and would allow utilities to bargain for the best services possible, and to take advantage of ever changing market conditions. After all, the real concern would not be with the market forces of reprocessing but with the disposition of the recovered plutonium and wastes.

With that point I now turn my attention to the international plutonium disposition regime or to be more precise, two separate but related regimes overseen by an International Plutonium Agency. One would see plutonium as an energy commodity that contributes to the energy security of member countries, and the other as a strategic material that for the benefit of all must be eliminated from the Earth. The link between the two would be the creation of a threshold amount permitted for energy needs; when exceeded, the plutonium must go to Space.

As an energy commodity, plutonium would first be used as MOX in power reactors (I exclude breeder reactors as their appearance is in the future). MOX, however, is harder to handle than fresh uranium fuels, since it is more radioactive, and it has different neutronic properties in a reactor. This means a reactor core cannot contain all MOX fuel, only a fraction. This is a complex subject, so for my argument here I will just assume it is one-quarter, that is, one fourth of a reactor's core would be MOX and the remaining three-fourths enriched uranium fuel. From this the threshold amount could be derived. To illustrate this, I'll use a very simple example. I'll assume a country wanted a 10-year plutonium energy reserve for its ten reactors. I'll further assume each reactor would require 500-pounds of plutonium annually. The total is easy to derive: 10x10x500 or 50,000-pounds or about 23 MT. So any plutonium from reprocessing under this amount would be legitimate, but any over must be designated for disposition. Of course, my example is technically wrong, but my point is not: This approach allows each country to determine its threshold for its energy security without others trying to impose their views. Yet it identifies what is excess and designates it for disposition. Moreover, this approach would not reopen the divisive and fruitless debate over reprocessing and MOX economics, as some countries would pay a premium for energy security. And it would not reopen the equally divisive and fruitless argument for mixing plutonium in a matrix material and burying it somewhere - unwise given plutonium's 23,000-year half-life. This says "Keep what you really need for energy security but dispose of the rest in Space."

In the real world though, some nations might not want a plutonium reserve, and some utilities might not want to use MOX, as it could lead to licensing questions, but both might want reprocessing as a waste management step. They would want to keep their spent fuel storage ponds empty, unlike U.S. utilities, which must find more and more space for their fuel because no Yucca Mountain has come into being in thirty years (and then face regulators who wonder why the public should pick up the tab). So they could designate all their plutonium for disposal, leaving only high-level and transuranic wastes to manage, both of which could also be disposed of in Space when heavy lifters appear. Under this new two-dimensional thinking, this might be called the once-through Space fuel cycle and become quite popular, as it probably would be cost effective and environmentally acceptable and leave only very low-level radioactive wastes to manage, like gloves and booties. After all, utilities are in the business to sell electricity, not to manage irradiated fuel, recovered plutonium, or high-level and transuranic wastes.

So the first step would be to create an International Plutonium Agency, a small single purpose organization separate from the multi-function IAEA and its worldwide membership. (The IAEA includes states with suspect nuclear programs, ones aimed at weapons development, violations of the IAEA charter and the international obligations they agreed to such as the NPT. This makes difficult the argument the IAEA should do it). The Agency would administer controls for the new regime that treats plutonium as an energy commodity, allowing it to be bought, sold, traded, bartered and exchanged in a market restricted to countries with robust nuclear power programs and exemplary credentials in the world of nations. In setting up this market, the controls regime, developed piecemeal over the past fifty years, would be upgraded to a single norm, and all member states would delegate to the Agency the rights to administer it. In a sense, it would resemble the agreement the United States has with Euratom but on a larger scale. (Euratom is a supranational organization of European countries formed to foster the growth of nuclear industries; the U.S. agreement with Euratom allows U.S. nuclear commodities to be exported to Euratom and moved about within those countries without further U.S. approvals). Countries with less robust nuclear programs or in regions of tension would be not only encouraged to join but also encouraged to sign contracts for commercial reprocessing services, and to begin transferring their irradiated fuels expeditious-

ly. Their requests to use MOX would be treated on a case-by-case basis, or they could be given an energy credit and provided enriched uranium in return, and they could be given preferential treatment in the plutonium launch queue. Some of these countries might welcome this option as it allows them to establish their bona fides with other nations while getting rid of their irradiated fuels completely.

How could it work? The Agency would be small yet highly professional and headquartered in Paris or London, but with offices in Washington, Tokyo, and perhaps Moscow. It would review, with a presumption for approval, contracts for reprocessing by its members, determining whether they were in accord with the overall regime. These contracts would include an estimate of the plutonium and other isotopes to be recovered. Contracts that present anomalies would be forwarded to the Plutonium Agency's Board of Governors, which would be composed of the diplomatic corps of the member countries for review and resolution. Once approved, the irradiated fuel enters the Agency's tracking and accounting system, a system that would apply to the plutonium throughout its stay on Earth. One part of this system would inform, on a confidential basis, all member countries of the amounts, forms, locations, and intended disposition of the plutonium. That intended for MOX would receive special attention, with appropriate authorities alerted in a timely manner as to when and how it begins its movement within the market, information that must be classified for security reasons. Other plutonium intended for an energy reserve would be compared against the threshold declared by the state: If under, no further action would be required; if over, it would enter the disposal queue. Plutonium withdrawn from the energy reserve would be compared with the amounts already in a country; if under the threshold, no further action would be required, but if over, it would raise red flags and be brought to the attention of the Board of Governors. Plutonium sold, bartered, traded, transferred, or exchanged among members would receive identical reviews. Plutonium designated for disposal or over a country's threshold would be mixed in a matrix material and poured in a payload container as quickly as technically feasible. Then it would remain at the reprocessing center until its launch time. After its disposal in Space, the Agency would adjust its books accordingly and make a public statement, alerting the world of the permanent disposition.

I expect NASA would conduct the launch, but as experience is gained, I expect NucRocCorp's division and commercial reprocessors would form joint ventures and pressure the Plutonium Agency's governments to enter the market. Why? It would be too lucrative to ignore. How? Small Re-core engines would carry out plutonium disposition, but different size heavy lifters might permit the economic disposal of high-level and transuranic wastes. So this regime might experience market pull

instead of government push. Those economics could only become better as different heavy lifters appeared and other engines on the continuum were developed. However, the pressure for privatization would be a decade-plus away, too long in the future to make any further comments here.

These bare essentials would require development and debate, provided that it was based on two-dimensional Earth-Space thinking. Already I see advantages. It comes to grip in a realistic way with a problem that will grow inexorably in the twenty-first century. By creating thresholds, it avoids the losing proposition of trying to impose abolitionist views of plutonium on others, particularly those who seek an energy reserve, yet it establishes a limit on the amount they may possess. This then focuses the spotlight on each country's threshold amount, and the prospect of continued pressure to lower it to zero. Anything over the threshold must go to Space, an activity that gives demonstrable and measurable evidence to the world of the success of nuclear disarmament and nonproliferation measures. Creating the International Plutonium Agency establishes a rewards regime for those countries with exemplary credentials and allows them to circulate plutonium as an energy commodity there. Yet a small, highly professional agency continually in contact with appropriate national authorities would administer that traffic, mitigating any security concerns. By linking this regime to the nuclear continuum, the appearance of increasingly more powerful engines would herald the emergence of the once-through <u>Space</u> fuel cycle, an activity with considerable environmental significance. It removes the oft-stated barrier to further nuclear power use, as the waste problem is solved, and allows increased reliance on it to help solve global warming, as nuclear plants emit no greenhouse gases. In other words, forget "cap and trade," that regrettable clause allowing companies to sell their carbon emissions quotas to others; this just shifts the location where carbon emissions enter the atmosphere, it does not eliminate them. If it causes global warming, then get rid of it; go to technologies with no carbon footprint, such as nuclear power.

Finally, this regime could create a powerful diplomatic tool to wield against those with questionable nuclear and missile programs, as they would be compelled to state their reasons for remaining outside. This would strip away any pretense they were for peaceful purposes, and leave them isolated in international forums. Moreover, their exclusion from NucRoc-Corp and the Space Charter Authority would create internal voices of dissatisfaction as they watched other countries and peoples prospering while they stagnated. In sum, I expect vigorous debate on the regime, but it must have accurate technical information. That is why I emphasized Congress must mandate disposition studies in the Democratization of Space Act. As better data emerged, Congress could then better determine the timing and specifics of a

Plutonium Disposition Act.

## III. Greening Our Planet: The New Thinking

I've raised this theme incrementally, so let me now combine it into a whole for the reader. In my Preface, I indicated nuclear rocket technology might be used to build reactors that melt the wastes clogging our landfills and contaminating our groundwater into a liquid that then is separated into the useful, the commercially valuable, and the wastes, now reduced in volume and in a form more suitable for long-term disposition, such as permanent disposal in Space. In chapter 4, I indicated some "free launches" might be Green, to test matrix materials for those wastes, or ideas that contribute to the goal of sustainable development, or solve global warming via space-based options. And I've just described a plutonium disposition regime. A common thread of the first two with plutonium disposition is the matrix material, which might be used in disposing of the melted wastes as well as other hazardous or toxic materials, which are difficult to dispose of, such as PCBs. Simply mix them with the matrix, pour it into a payload container to harden, then launch it to the Sun, which will annihilate it.

Herein lies one part of the new thinking: With continually improving heavy lifters, the disposal of hazardous materials or wastes or other toxic debris of human civilization in Space becomes economically feasible. It's inevitable, it's inherent in the technology, the same way the lunar landings were inherent in Goddard's rocketry and the Nautilus's feats were inherent in nuclear submarine technology. So to facilitate this new thinking, when Congress passes the Democratization of Space Act, it should mandate DOE, NASA, and EPA to identify these substances and the facilities needed to process them, develop flight-rated payload containers and matrix materials, then conduct a launch program to prove the technology to the private sector. In other words, demonstrate the capability, show the private sector how it can be done, but let it takeover. Let it build the processing facilities and payload containers, let it produce the matrix materials, and let it form partnerships or sign lease agreements with NucRocCorp's divisions for the launch. Let all make a profit. This creates good-paying jobs and expands the tax base, not only in the United States but also in other regime member countries, those eligible to use nuclear rockets. Obviously, this new capability would require its own international treaty – the Outer Space treaty does not cover it - so that would have to be negotiated, but the DOE, NASA, and EPA demonstration program would give that effort critical data.

Just this capability *alone* would add a new option to the Green's arsenal: Where previously they used demonstrations, persuasion, and lawsuits to achieve their goals, now they could incorporate as a for-profit business to form disposal companies. In other words, they could now have mar-

ket pull working for them, and they should have no problem entering into a lease or partnership arrangement with NucRocCorp and its divisions. Actually, I expect some of those divisions would not need prompting, eye this market closely, and probably come to the Greens with joint venture proposals.

The other part of the new thinking lies in what was learned from the Green "free launches." In addition to matrix materials, there might be three main areas: Better space-based sensors for environmental monitoring; promising space-based solutions to mitigate or eliminate global warming; and solar station designs to beam power back to Earth. The latter two, of course, would require "demonstration and proving," but it need not happen in the "put up or shut up" or "convincing stations." In other words, heavy lifters offer the environmental movement the opportunity to construct its own space station. I would see no problem in obtaining a charter - this worldwide movement should have no difficulty in raising the necessary capital - and I see every incentive for them to do so, as having a Green space station would crystallize their movement internationally, and allow them to set their agenda on how it is operated. For example, they could test solar power concepts, offer space-based Eco-tourism, host conferences, provide ecological training, monitor the Earth, or broadcast back to Earth; no doubt, as this new thinking took root, many other uses would be found. Furthermore, its membership would be an excellent source of funding to commercialize any technology demonstrated in the Green station, and begin making a profit. So breaking the taboo really puts all alternatives on the table for solving Earth's environmental and climate problems, not just some, and it forces those concerned with such matters to think and act in Earth and Space dimensions, giving them more solutions than they currently have, and probably profitable ones as well.

## IV. The Russian Option

Any international plutonium regime must have Russian participation- even though the West is reviewing its relations with Russia after its recent return to thuggish-ness by its rollback of democracy, assassinations, invasion of Georgia, and bellicose warnings of a new Cold War. This option should be part of that review, as it offers the opportunity to create three countervailing trends: Establishing for-profit industries, introducing democratic forces deep in Russia, and elevating the Duma, the Russian legislature, as a check on the executive branch. This would be a variation of Niels Bohr's thought of nuclear rockets permitting global nuclear disarmament, and it would feature the unconventional thought of legislators exercising dual sovereignty. All right, how?

In central Russia lies the Krasnoyarsk nuclear site that had been a vital part of the Soviet Union's nuclear weapons complex. It featured dual-purpose nuclear reactors (producing electricity and plutonium) and reprocessing plants dug deep into the mountains, hidden from U.S. spy satellites and secure from conventional air strikes. In the 1980s, the Soviets began constructing the RT-2, a gigantic aboveground reprocessing plant, to be larger than those of the British and French, to handle the irradiated fuel from its future light water reactor program. However, the demise of the Soviet Union cancelled both plans, as no money existed to complete them. My proposal then is simple: Complete this plant and open it up for commercial reprocessing, including U.S. irradiated fuel.

In doing this, I rule out certain options quickly. First, Russia does not have the public money to do it; to try it would raise storms of protest when pensions are paid chaotically, and this is why Russia tried unsuccessfully to find Western backing in the 1990s. Moreover, the current Russian government appears to be dedicating all its oil revenues to financing foreign policy gambits with its military. Second, the private sector in the United States or elsewhere would be unwilling to finance it given the unfavorable attitude the Russian executive branch has toward private companies, and their corporate officers would be unwilling to pay-off those officials or endure harassment and intimidation.

There is another option, however, the Duma, the Russian legislature. I propose the Duma create a special legislative district in Krasnoyarsk, carving out the reprocessing plant and an adjacent area of perhaps 100 square miles, or whatever is reasonable. It would contain everything required to complete and operate RT-2 as a commercial endeavor. That would include housing, banking and financial institutions, schools, shopping and cultural activities, media and communications, and a local government with legal and security institutions independent from the Russian executive branch. In large measure, this would be a closed city reminiscent of those in the U.S., such as Oak Ridge and Los Alamos during World War II, and those that the Soviet Union had around its nuclear sites. So this would not be new. What would be, however, is that U.S. and Russian legislators would exercise joint sovereignty over it for a specified period, perhaps 50-75 years, under a joint management corporation with carefully delineated checks and balances. In other words, the Russian executive branch would have no authority over it.

The reprocessing complex would be jointly managed, though at the beginning I expect the United States would have the lead to finish the plant to high reliability, safety, and environmental standards. At the same time, Russians would be sent to American schools or organizations to develop the skills of running a large enterprise for a profit. So over time, they would assume leadership and high-level management positions in the complex. All recovered plutonium would be disposed of in Space and all high-level and transuranic wastes as well. The corporation would also have the right

to reprocess irradiated fuels from other countries on a for-profit basis with identical treatment for the plutonium and wastes. And the corporation would have sole authority over the complex, including its on-site security apparatus, with a board of directors to oversee and judge any differences with day-to-day operations. And it must have unfettered port and rail access from both coasts to transport the spent fuel to Krasnoyarsk and recovered plutonium and wastes from it, and unfettered air, highway or rail access for U.S. and Russian personnel working at the site. A U.S.-Russian legislative commission would appoint board members, perhaps three or four from each country, as well carefully delineate the corporation's charter; one feature of it would be its financing. Part of the residual tax money the U.S. utilities paid for the once-through fuel cycle could be dedicated to completing the complex, with the remainder coming from the private sector. This could come from US utilities seeking preferential terms for contracting early, U.S. and Russian banks that see the profit potential of the special legislative district, and Russian and U.S. taxpayers both of whom should be allowed to participate. Finally, I expect this company would be listed on the stock exchanges in the United States, Russia and elsewhere.

This U.S.-Russian legislative commission would also oversee a council elected by those within the district to administer local affairs, including for-profit businesses. I expect this would be quite profitable as modern housing, shopping centers, schools, and so forth would be needed. This would be important to attract the highly skilled American workers needed in the complex, and their Russian counterparts would expect equal treatment. The legal and security institutions would have equal numbers of Americans and Russians but alternating leadership. For example, the police chief would be a Russian for two years, with an American deputy, while an American would head the court, with a Russian deputy. Command would rotate for the next two years. This maintains fairness and fosters a spirit of cooperation as neither could unilaterally dominate for longer than two years. This should prevent the appearance or fact of one country dominating and giving unfair treatment. A check and balance would exist in every institution.

This U.S.-Russian legislative commission would meet at least once every five years to review and make appropriate changes in the corporation or district. Its membership could be three or four from each country, all with equal voting rights. I thought long and hard about having an odd number as that creates the tie-breaking vote in democracies. However, my instinct says an even number would be better, as it emphasizes compromise and consensus building in the decision-making process. I expect the legislatures of either country would select as members those who have the utmost personal integrity and are adept in the democratic arts of compromise and consensus building. I want a pragmatic commission that never loses sight of the ultimate goal of reducing inventories of plutonium, not one that becomes caught up in the pursuit of personal gain, or ambition, or winning. Finally, I would stipulate that if the commission were ever unable to reach consensus, all must resign and never be allowed to return, with each legislature appointing new members. This further emphasizes compromise and consensus building.

Some might say this is a risky scheme and shouldn't be pursued, but I say one-dimensional thinking would be even riskier. Here we could not only set up a privately financed means to reprocess U.S. irradiated fuels, marking all plutonium for permanent disposal, but also introduce democracy, capitalism, and a respect for law deep into Russia while at the same time promoting the Duma as a check on an unrestrained executive branch. If workable, this model could influence legislatures in other troubled regions of the world to create special legislative districts, such as Kashmir, a continuing flash point for India and Pakistan.

### Postscript

In July 2007, Presidents Bush and Putin initialed an agreement to cooperate in the peaceful uses of atomic energy. The Congress and Duma both must approve it. Following Russia's invasion of Georgia, the United States withdrew the agreement, but it is not clear whether that is temporary or permanent. The agreement itself seems to provide exporting U.S. irradiated fuels to Russia for long-term storage, perhaps with reprocessing in the future. This isn't clear, and seems to imply the once-through fuel cycle is defunct in the U.S.; in other words, Yucca Mountain is dead. If so, this is at best a halfway "out of sight, out of mind" measure. It merely shifts U.S. spent fuel to Russia, and postpones a decision on its reprocessing. It doesn't cover the plutonium growth and disposition problem in a global sense, as it is only a bilateral agreement; and it provides no role for the Congress and Duma other than approving the agreements.

---

[i] Heavy water reactors use natural uranium processed directly into fuel elements, thus avoiding the enrichment step, but their irradiated fuel requires treatment the same as for other reactors.

# Chapter 7
## Four Political Structures of Democratic Space

*This chapter discusses the Nuclear Rocket Development and Operation Corporation and how the Space Charter Authority and Space Regulatory Commission derive from it. It also inquires into the Space Council/Space Assembly that would have oversight over them. Lastly, it considers the role of the stockholding public to see how it could provide oversight to all the above, transforming the space program into a truly democratic one.*

## Introduction

I have discussed the need for developing an infrastructure to maximize the nuclear rocket's potential, introduced NucRocCorp and the Space Charter Authority, and alluded to the Space Regulatory Commission. Here I analyze them in greater detail and inquire into the Space Council and Space Assembly, the final structures necessary for a truly democratic space program. All four would foster private-sector development of Space, creating a worldwide, tax-paying space industry to construct man-made objects in a vast, hostile area without boundaries. That would make it different from the nation-state, which ultimately is concerned with boundaries, often defined by oceans, rivers, and mountains, and sovereign rights within those boundaries. This would be a new regime, and membership in it must be exclusive to countries sharing certain values. It must also be voluntary - no one would be forced to join or participate – and fraternal, rewarding good behavior and punishing bad. It would not recognize claims of territory or birth as having any validity. In other words, "one man,

one vote" would be inapplicable. That means a citizen of a nation-state has a right to vote in its political system, with all votes being equal – no one with more votes than another. Instead, I advocate proportional voting, where the votes a person has, equals the shares he or she has. A thousand shares, a thousand votes; no shares no votes. After all, why should those who decline to participate in a voluntary for-profit regime have a right to determine its governance?

## I. Structure of the Nuclear Rocket Development and Operation Corporation

I begin with NucRocCorp, first created to develop the small Re-core and Re-use engines, but it would be a transitional organization, that is, certain responsibilities would transfer to the Space Charter Authority, Space Regulatory Commission, and Space Council/Space Assembly after the decision to build heavy lifters. So it would develop, operate, launch, and recover small engines, and use that experience as a test-bed to debate the functions of the other political structures. Thus, some NucRocCorp structures would oversee NucRocCorp and its divisions *and* prepare for their newer and larger roles. For them, one eye would be on the present, the other on the future. (Here, I refer the reader to the NucRocCorp organization chart in Chapter 4; I introduce the transformed model shortly). After this transformation, NucRocCorp would still exist, but only developing and operating more powerful engines, and building payloads or space stations. Initially, it would have three structures: The corporation proper, divisions, and boards of directors.

### A. NucRocCorp's Division Structure: Development, Operation, Launch and Recovery

Within NucRocCorp might be any number of engine development and payload and operating divisions – the market would decide - but just one division to launch and recover engines. Thus, the development and payload/operating divisions could or could not be linked. For example, some companies might be satisfied to develop engines and a stage, then lease or sell them to others to operate, probably providing an engine re-building and re-certification service here as well. Others might want to make the payload also, thus making use of the parent company's full range of product services, and operate it for the customer. Or still others might build the payload, specializing perhaps in certain missions or profitable niche markets, and leave engine development

and operation to others. After heavy lifter development begins, all might reorganize and specialize in space stations. How each organizes its involvement is a business decision.

Each division would determine its own structure, its management, labor, pay scales, and officers, including any who exercise oversight. Those could come from the parent company or elsewhere. Each division would operate under its own business plan, price its products and services accordingly, and keep a preponderant share of its profits, using them in accord with its business plan (i.e., reinvest, dividends, expansion). The remaining launch-and-recovery division would be all-government initially, but be privatized within a decade and with several new ones probably coming into being to compete. They would operate with a government license and have their own officers, business plan, etc. Neither the government nor the public would have any direct involvement or role in running any division - about the same as it is for other corporations under capitalism. After purchasing at least one share in NucRocCorp, the taxpayers could purchase as many shares as he or she wants in any of the divisions, or to fund any mission they propose.

### B. NucRocCorp's Structure and Oversight of the Divisions

A separate entity, NucRocCorp would have management oversight of all divisions; it could be similar to the structure of General Motors, and its competing divisions, such as Chevrolet, Saturn, Pontiac, Buick, and Cadillac. Here each division competes in the market against other automobile companies as well as other divisions in General Motors. Yet they cooperate with each other, using the same parts and so forth. I expect something similar here where each division sees competition and cooperation as the key to greater profits. But there is a difference. Each division would nominate candidates to serve in NucRocCorp, such as president and chief financial officer, on the basis of technical competence, business acumen, or management talents without any shareholder influence. However, only NucRocCorp shareholders should vote them into office on the basis of one share-one vote or two share-two votes. This creates a check and balance. I like long tenures of office to create stability at the top, something important for business as frequent changes create doubt and uncertainty and affect performance. I also like periodic shareholder oversight and the removal of underperforming executives. Hopefully, further debate would clarify the tension between the two positions.

NucRocCorp would have many functions; I list ten key ones to begin a dialogue.

First, it would fund and manage fuel element and advanced nuclear propulsion research and development. Both would require close integration with the engine development and operating divisions, and data sharing committee (below).

Second, it must develop the solar system transportation plan, in cooperation with NASA and other space agencies, and SSTO design criteria, and publish both.

Third, it must fund both "garage" design studies, and at some point in small engine development, perhaps by the second-generation one, must approve development and build both. This would be a cooperative effort with NASA, the space agencies of other countries, the divisions, and the private sector, as the goal would be to create a space station industry in member countries. NucRocCorp's divisions and stockholders probably would fund this to have control over patents and intellectual property, and this would be very profitable as the decades pass by. If it's government funded, it becomes public domain, and NucRocCorp stockholders get no return on their tax dollars.

Fourth, it must develop criteria for creating new divisions, foremost of which is the "entrance fee" they must pay to join. This should be substantial, and include an even bigger "compensation fee" for those who join and start a division after a certain date. A "Johnny-come-lately" should pay for the prior work, and the founding divisions compensated appropriately. In other words, NucRocCorp and its divisions would be market driven, and free "piggy-back" rides would not be allowed.

Fifth, with this entrance fee, it must

design, build, and operate test facilities, scheduling their availability in a fair and equitable way for each division.

Sixth, it must translate policy guidance from the boards of directors into management directives for all divisions, including guidance from NASA and DOE. Linked to this, it must ensure compliance with the directives.

Seventh, it must levy a "division tax," a sum each division must pay to help fund NucRocCorp. This probably would be light at first until experience was gained, but increase thereafter so NucRocCorp could pay a dividend to its shareholders.

Eighth, it must conduct aggressive technology outreach efforts for marketable innovations, patents, and know-how of its fuel fabricators and advanced propulsions experts, and reactors developed for special applications, such as process heat. This could include those of the divisions, or they could do so on their own.

Ninth, it must conduct vigorous public relations campaigns not only in the United States but all member countries, telling citizens how they could prosper from their investment or from their ideas about using Space.

Of all this, however, the tenth is the most important: NucRocCorp must decide when to develop the heavy lifter, and it must use only technical criteria. I expect it when fourth-generation small engines appear, but the

decision must not be left to any others, particularly those in the political realm. Why? Building heavy lifters signify the political infrastructure to govern their use must be created. Political leaders tend to debate and postpone difficult decisions; giving NucRocCorp this authority tells the legislatures of the member countries they could debate the structures up to then. Afterwards, they must create them, and NucRocCorp investors must ratify them. If the political realm could not do this, NucRocCorp must have default plans for the institutions to take to shareholders, in effect, by-passing the legislatures and allowing shareholders not to be held captive by the political process. Let them approve if politicians could not. In this space program, shareholders are sovereign. I'll say more about this later.

Obviously, the structure and personnel NucRocCorp require to perform these functions need debate. Since some divisions might be foreign owned and operated, NucRocCorp must ensure its officials come from all member countries, with all having meaningful responsibilities, but featherbedding and tokenism must be avoided. This ensures NucRocCorp has a broad outlook with technically competent and business-wise people to establish the practices and procedures the divisions would follow.

## C. The DOE and NASA Presence in NucRocCorp

DOE and NASA must have a permanent presence in NucRocCorp,

since test and development facilities would be on DOE sites, since DOE would provide the uranium, since NASA sites might be involved in small engine development or as a staging area for cargo plane launches, and since NASA and other agencies would purchase nuclear engines for government use at cost. So both must have clear and unambiguous authority to intervene in any or all of NucRocCorp or its division's activities on matters concerning the sites, such as environmental concerns and security of engines in storage, transport to launch and transport back to DOE sites for reprocessing. This authority, however, must not extend to any division's technical development, launch and operational activities, and business affairs. Obviously, DOE and NASA would determine the nature and extent of the representation and involvement, no one else.

However, both agencies would have larger and more important roles. Both would be developing plutonium matrix materials and certifying containers, a process likely to involve flights of the first-through-third-generation small engines. Then Congress would restructure both to support the "free launches" by making it easier for the public, the "moms and pops," to access either's laboratories or facilities before or after the launch to bring their ideas to economic viability. That would include assisting NucRocCorp and its divisions, or the private sector in the design and development of the "put up or shut up" and "convincing

stations," but here with the thought to create a space station industry, like the assistance NASA and DOE give to the airline and nuclear power industry. Indeed, if NASA has giant wind tunnels to test the aerodynamics of flying objects, it might need similar facilities to test components to support a burgeoning space station industry; DOE might need reactors to test space station materials in a radioactive environment. And Congress might need to restructure both to streamline their costly and time-consuming contracting procedures that deaden their interactions with the private sector today. Also, both would cooperate with the EPA on the disposal of hazardous materials in Space, on matrix materials and payload containers, and then on demonstration flights. Finally, this restructuring would include research and development leading to a space regulatory regime, e.g., space station fire suppression standards.

## D. Fuel Development and Advanced Concepts in NucRocCorp

I have placed fuel development and advanced concepts within NucRocCorp because I expect those to be composed of people from NASA, DOE, other U.S. and foreign laboratories, and conducted on DOE sites where strict security measures are in force. However, this is debatable. Industry has groups that push the state of the art in technology or develop advanced concepts, so each engine development division could carry out these functions. Further discussions would clarify this, but I suspect it would become unattractive, as it could dilute the engine effort, and incur more costs than an individual division would like. One thing, however, should be uncontested, that *at least three* different groups must be involved in fuel development, each with its unfettered ability to test fuels and materials on its own schedule, yet with each sharing its results with the others. This must emulate Rover/NERVA, where Los Alamos, Y-12, and Westinghouse engaged in spirited yet friendly competition, and the program benefited immensely. I suspect many DOE sites would be keen to host this.

### E. Public Accounting

NucRocCorp must have a permanent public accounting office, as money would flow into it from three sources. First, if other sources fail, government must fund small engine development, and this would be "get the ball rolling money." It must be repaid. The second source would be the "entrance fee" the private sector must pay to join. Though undetermined now, it could be something like $500 million each and if three, four, or more firms joined, that's several billion dollars and that could fund much if not all of the technical infrastructure, making the government's funding obligation perhaps non-existent, just a contribution in kind. The private sector would have incentive to join early and avoid the "Johnny-come-lately" compensation fee. The third source would be from citizens via their tax forms, and this might be inconsequential at first but become significant as the first small engines begin flying, particularly with "free launches."

So three revenue streams would flow into NucRocCorp and its divisions. Government totals would be important in determining the tax revenue each country should receive, as developing nuclear rockets means developing a broad-based, worldwide, privately funded space program, and at some point governments must tax it. However, a substantial tax base might not exist until heavy lifters and a space station industry appear; nonetheless, those governments who got the ball rolling should be repaid for their contribution *and,* perhaps, receive proportionally more tax revenue than those who invested less. I'll discuss this more in the next chapter. Finally, keeping track of privately invested money would be important in allocating dividends.

So those in the public accounting office must have utmost integrity, and the unimpeded authority to look at the books of any division suspected of fraud and wrongdoing. They would keep records of the contributions and pay dividends, and should have the final say about the compensation package of NucRocCorp officials, but not the divisions. Just as fraud and wrongdoing could stifle the public's willingness to invest in Space, so could bloated salaries and excessive retirement and benefits packages.

## E. The Data Committee

Finally, NucRocCorp would have a data committee, with its members selected by the divisions with no NucRocCorp, government, or public involvement. It must ensure that data obtained from flight and operational experience was shared among all divisions, but this would be sensitive, as it would involve propriety data, patents, inventions, and know-how. In other words, some divisions might make better performing, more economical engines that others, and so would be unwilling to share their expertise with competitors. Yet the overall object must be to transform the space program into a privately funded one, with safety being of paramount importance. This committee, therefore, must develop criteria for the data sharing, and then ensure it was. I expect safety would be on the top of the list, but its definition could encompass all operations so it could become meaningless. Still, I think any concerns the divisions might have here would be more perceived than real, as industry is adept at striking technology-sharing deals that benefit all parties. I expect something similar would occur here, particularly as each division realizes the way to increase its profit is to make the pie bigger, and not wrangle over the size of a slice from a smaller one, as is the fate of the chemical rocket industry today.

## F. NucRocCorp's Boards of Directors

I mentioned repeatedly NucRocCorp would have five boards of directors, and not one with five different functions. That would be too limiting to innovative thinking, and hinder the transformation of four of them into the institutions that would govern heavy lifters. These four are those about which I said one eye must be on the present and the other on the future. I hope to have a trained cadre that could lead when the time comes.

1. <u>A directors' board</u> would oversee the divisions. Countries with divisions developing or operating nuclear engines should be on this board, perhaps one per country, perhaps with a limit of ten total. Each should nominate several candidates, and from this list six would be elected; the remaining four should be elected from a list of candidates developed by stockholders. This board would remain after the others are transformed.

2. <u>A business board</u> would oversee business operations of the divisions operating the small engines, righting any unfair practices, and ensuring all NucRocCorp member countries had access to flights. Its more important task, however, would be to draft charter terms, ensuring all member countries had an opportunity to comment on them before it was transformed into the Space Charter Authority and actually issued them. It would work with the

Public Accounting Office to develop a worldwide financial system to handle money invested in the charters.

3. A membership board would oversee who could join NucRocCorp and by implication the Space Charter Authority and other political institutions. I discuss this further in the next chapter, but let me state briefly all countries joining NucRocCorp would appoint one member for this board, with an equal number elected by stockholders. That would make this a potentially large and ineffective group, yet it would be important. It would also determine criteria for membership and decertification, and make recommendations on applications to join as well as punishments for violations. Any official on this board must have the right to call for a vote on these matters. Keeping it a mix of government and stockholders would ensure wayward actions by members, or their peoples, could be called to account quickly and not glossed over, as all too often happens in the United Nations. Finally, with the business board, it would also develop interim criteria for the first zone and this would be limited to LEO initially. This board would be transformed into the Space Assembly.

4. A regulatory board would begin development of a regulatory regime. It should be composed of one representative from countries having engine development or operations divisions. The exact numbers of officials and their tenures would need to be deter-mined as experience was gained with the small engines, but all serving here must have technical expertise and a regulatory mind-set to ensure safety, with profit at the bottom of the list of priorities. This group should have different working groups - pre-launch, launch, space operations, recovery – and it must have authority to direct NucRocCorp to carry out research and development or studies either in the divisions or elsewhere. Its work should be public to the extent possible, and its decisions not overturned or disallowed by the Steering Committee (below) unless overriding circumstances so warrant. So a check and balance would exist. It would be transformed into the Space Regulatory Commission.

5. The Steering Committee would be the fifth board and oversee the others in the small engine period. It would be composed of high-level government officials nominated by the principal countries in NucRocCorp. Stockholders would not elect them. It could include the heads of each country's space agency or whomever each country's legislature designated, and their terms should be five years in length. It would approve new admissions as well as any disciplinary actions; it would overturn any regulatory actions deemed excessive or unwise, or remand them for further work; and it would initiate new policies and procedures. However, its relationship to the Public Accounting Office needs debate. I like the idea of oversight, but I deeply dislike the idea of government

officials having any control on this private money, other than to ensure that governments were repaid for their "get the ball rolling money," if that was necessary. I also dislike the idea of their having any review of compensation and benefits packages.

The Steering Committee's most important function, however, would be to review NucRocCorp's decision to build heavy lifters. After consulting with their respective legislatures, it could formally ratify or remand the decision, two choices only. No postponements, no delays. If the Steering Committee ratifies, the plans must be submitted to stockholders for their ratification. They, the citizen stockholders in the member countries must have final approval authority in this new regime, not legislatures. If remanded, NucRocCorp must have default plans for these institutions, and submit them to the stockholders for their vote. If stockholders approve, then the heavy lifter development must begin, and the Steering Committee must oversee the transformation of the boards into the new structures, a process taking about a decade.

Either way, that decision would set in motion a chain of events leading to the appearance of the Space Council and Space Assembly, the Space Charter Authority and Space Regulatory Commission. During this interim period, the Steering Committee would have four principal tasks. First, it would authorize development of the "put up or shut up" and the "convinc-

ing stations" and the bus and nuclear power reactor. NucRocCorp should not have that authority, as the intent has been all along to build a competitive space station industry for Zone 1 colonization, and the Steering Committee should be the primary institution to experience its growing pains. I expect funding would be easy to obtain, as the decision to build heavy lifters would be widely publicized, and all would know a space station industry is to follow. Second, it would task the business board to publish the charter terms and conditions, giving the private sector time to understand them. This business board would then revise them as appropriate and publish them again. Third, it would task the membership board to publish the criteria for membership in the Space Charter Authority and other political structures. That should have been debated for years, but it's important now as many more countries are likely to join this regime than NucRocCorp. They must know that membership is not permanent like with the UN, but based on good behavior by the government and its people. Fourth, it would task the regulatory board to transform itself into the Space Regulatory Commission, and be deeply involved in the design and development of the "put up or shut up" and "convincing stations," bus, and nuclear power plant. Safety must be built into these structures from the beginning. Finally, the Steering Committee itself must prepare to be transformed into the Space Council.

## Membership Incentives

When NucRocCorp is created, a window of opportunity should exist for other countries to join without cost. A similar window should be open for their companies to join or enter into partnerships, though they would have to pay an "entrance fee," the same as their American counterparts. After NucRocCorp has operated for several years and gone through several generations of small engines, countries wishing to join should pay a fee, though its size would need debate. Now membership should not be free as before to make clear to all this is to be a privately funded, for-profit space regime, not a government largesse effort like the one that has characterized the space programs of all countries since 1957. So some fee should be established for countries and an even bigger one for firms seeking to join NucRocCorp. They should "compensate" the divisions who began the work. Both fees should be even larger when the "put up and shut up" and "convincing stations" begin operations, but the money here should go to the Space Charter Authority to pay down the bond it floated for them. Structuring membership in this fashion should be a big incentive to encourage early membership.

## II. Structures of the New Space Regime

Two-plus decades might have passed since NucRocCorp was formed and began small-large engine development, and I would be presumptuous to assume I can state what comes into being. I can, however, offer thoughts to begin a debate. First, I would expect the regime would have more members than those in NucRocCorp, creating a problem of giving all a role in governing the colonization of Space. New political structures, the Space Council and Space Assembly, would be created to solve this.

In considering these, I looked to the UN's structure for guidance: It has more than 180 members, who are divided into a General Assembly and Security Council, the latter with fifteen members, five with permanent seats as they hold the most power in the world. The General Assembly, composed of all other countries, elects the other ten members for two-year terms. Security Council decisions require nine votes, but any permanent member can veto a decision. The General Assembly admits new members, approves the UN's budget, and establishes agencies to carry out its mission to maintain peace and promote cooperation in solving the world's problems and upholding human rights.

Second, I expect a different set of circumstances would exist several decades hence, and a different human psyche. Why and how? The solar system transportation plan would probably be fine-tuned by then, with routine runs to the planets. Vast improvements to the solid core would be likely, with many different types flying and with capabilities well beyond fourth-generation systems. Perhaps other engine technologies on the nuclear continuum might be in service or in the develop-

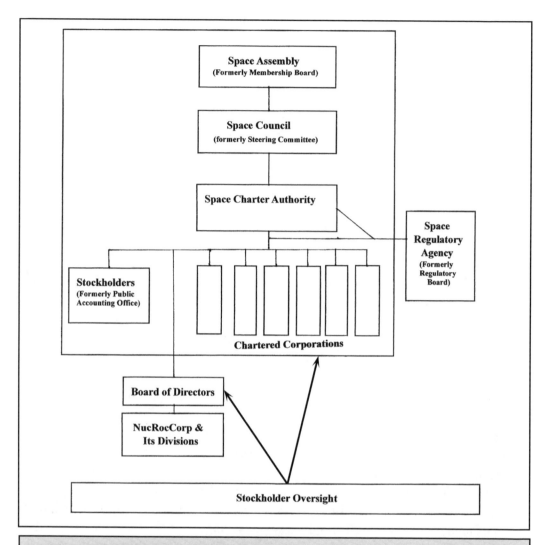

**MANAGEMENT STRUCTURE DIAGRAM**
 This organization chart illustrates the four political structures necessary to govern heavy lifters and the private sector colonization of the solar system they allow.

ment pipeline by then, promising even greater reductions in the cost of moving payloads into and around Space. Then the "free launches" would continue to have their effect, and people everywhere would see how they could benefit economically from low-cost access to Space. This would include the Greens, who might have formed their own launch divisions to dispose of hazardous materials. Their own space station perhaps could be looming. Citizens would probably be flying to the "put up or shut up" and "convincing stations." A burgeoning space station industry would be hiring furiously and planning for launches. Finally, all of this might prompt debate about opening up new zones for colonization; the Moon could be within economic reach. In sum, all this might repeat the 1950s where Space was ever-present in the U.S. psyche though now it would be a global awareness, and one attuned to an individual's newfound right of dominion over the solar system.

## A. The Space Council

With that as background, my thoughts begin with a permanent group of space-faring nations, that is, those countries with substantial space programs that meet the membership criteria. I would expect its number to be small, the United States, Canada, and Japan, and Europe entering either as a bloc or as individual countries; hence, it would be similar to the UN's Security Council. Actual membership would have to be determined. Russia may or may not be eligible, and based on its current behavior, I suspect the latter. The Steering Committee would evolve into this Security Council-like group, but its name is a problem. It should signify its purpose and it would not be security, but private sector development of Space under government auspices. Though not a good name, I'll just call it the Space Council. Its responsibilities would be different from the UN's Security Council, since it would be concerned with promoting private sector colonization of the solar system, and resolving problems uncovered in so doing. One function would be to ratify opening of new zones and also approving recommendations from the Space Assembly. Each legislature of member countries should nominate candidates to serve on the Space Council, but from this slate, stockholders would vote some, but not all, into office. Moreover, stockholders must have oversight over the Space Council, with the authority to overturn or remand its decisions.

## B. The Space Assembly

An entity like the UN's General Assembly would assume functions of several boards; it would evolve from the membership board and be responsible for establishing criteria for zones, membership, and disciplinary measures. It would also have some responsibility over charters, where it could propose changes to them, and it could create a legal system to resolve disputes with them. I don't have a good name for it, so I'll just call it the Space Assembly. Each country joining the regime would nominate a member to serve and each would have a single vote. The role of stockholders here is debatable and many options exist, from allowing them no role whatsoever, thus keeping the Assembly purely a government tool, to allowing them to elect an equal number of members, thus bringing universal suffrage to Space and creating a check on governments. The reason why is simple: Among its duties, the Assembly would have power to tax those structures built in Space to fund the political institutions, and taxation without representation will resonate in Space as it resonates on Earth. So here an option might be equal representation: If the Assembly had 100 from governments, stockholders would likewise have 100, making the Space Assembly composed of 200. Thus, many options exist for debate.

The Assembly should have authority to pass legislation by a simple majority concerning the charters and chartered, including the aforementioned taxes; to establish budgets for the Space Charter Authority and Space Regulatory Commission; to hold public hearings and investigations, and recommend new zones for development. The latter would be particularly important, as it would allow all member countries a say in the colonization of the solar system. It would not just be the industrialized ones flexing their muscles. In many respects, it would have functions identical to legislatures everywhere.

The Space Council and Space Assembly structures seem top heavy and cumbersome for so little Space activity at the beginning, even with the "put up or shut up" and "convincing stations" flying - ready to tax and discuss at a moment's notice - thus slowing the expansion into Space. Perhaps that might happen, but I see a transition period where the Space Council and Space Charter Authority act alone until other space stations appear. So when the first space hotel or manufacturing station starts operations, when more powerful engines appear, the Assembly begins its work. In other words, I see steep bell curve growth, with large commercial space stations appearing with increasing frequency after the "put up or shut up" and "convincing stations." This creates a large tax base in LEO. How much and how fast is

now uncertain. I know many would scoff at this, but they must not ignore basic economic. Going to LEO for $30 million per ticket means, as we know, extremely low demand; going there for $87 a pound or much less would mean extremely high demand. It's Economics 101. Indeed, already many have signed up for the "roller coaster" suborbital space ride at $250,000 per ticket.

The SSTO could change some of this as the politics of establishing spaceports in Latin America and Africa and emergency landing strips in the Pacific islands would dominate, particularly as governments jockey for position to host these facilities. This could affect the creation of these institutions and spill over into the UN; that only complicates the politics of creating them. It does not eliminate the need for them. For example the SSTO would still have to fly to something there; that would be chartered space stations; and as the Assembly opened more zones, such as geosynchronous orbit or the moon, other spaceships would be needed to go there, as SSTOs would end their flight in LEO.

### C. Space Regulatory Commission

I don't know what the Commission would look like, but I hope to make some observations to begin debate. First and foremost, I find it inconceivable to have private sector development of Space without a strong regulatory presence. What works for a chemically propelled space program, mostly run by governments, could not work here. Second, the Commission's origins would be in NucRocCorp's regulatory board; only the space-faring countries on the Space Council should have their nationals in positions of leadership in here, with little or no consultation with the Assembly or stockholders. Third, the number of commissioners requires discussion; there should be more than one to allow different viewpoints but less than ten to avoid a clumsy, ineffective structure. My best guess is an odd number of five, seven or nine, each of which would receive a five-year term, each of which would be staggered to permit continuity and change, each of which could be re-nominated to another term, and each would have a different technical expertise. Such could include nuclear physics, structural engineering, electrical engineering, aerodynamics, nutrition and medicine, and industrial and fire safety. Fourth, a majority vote would rule, but how the Commission would develop its rule-making procedure requires debate, particularly on the comment and appeal process. The Space Council should be the court of last resort in the appeal process, but that would draw into question its competence on technical matters. Nonetheless, this requires thorough examination, as not all rules result in a safer product and too many could actually hinder safety, as it might overemphasize the system and not the people operating it. With hundreds on board each space station, failure is not an option.

## D. The Space Charter Authority

Deriving from the business board, the Space Charter Authority must be market-driven, and so its personnel must promote space development. Hence, it should have wide latitude in its operations, including the right to issue and re-issue charters without further review by anyone, to adjust to market forces and to operate the "put up or shut up" and "convincing stations" to promote the development of a space station industry. Other powers would include making revisions to charters, establishing promotional incentives such as lotteries, and perhaps creating worker educational and training programs. And it would translate the approval of a new zone into nuts and bolts reality, perhaps by approving only certain types of charters initially to ensure the opening of a new zone was done safely. So the Authority should be adept at re-evaluating its fee structure and operations and listening to its critics as well as supporters, a process that would become institutionalized when the Space Assembly appeared. The Authority's testimony here would be frequent. To do all this, the Authority must have innovative thinking at the top, yet it, as with most corporations, must have one person in charge to keep the organization focused on achieving its goals. A Commission structure is wrong here; what is appropriate for regulation is not for promotion. Too many cooks and so forth.

Yet the Authority's leadership must be subject to oversight. One way could be for a director and deputy to be nominated by the Assembly and confirmed by the Council. Both should have a business, not government, background. This could have a North-South orientation, a director from the industrialized countries and a deputy from the less industrialized, to satisfy governments represented in the Assembly and space-faring countries on the Council. But it might be unacceptable to stockholders. To satisfy them, the offices of general manager and chief financial officer could be created, the former to oversee the Authority's day-to-day operations, and the latter its disbursement of money. If only stockholders nominated and elected these, the Authority would have a check and balance within itself, allaying concerns the Assembly-nominated officials might misuse money on schemes of dubious worth or on proposed charters of doubtful merit. On the terms of office, I like continuity at the top, suggesting long tenures, but I also like oversight and removal of underperforming executives. Debate should clarify this.

### III. The Role of the Stock Holding Public – The Stockholder's Institution

In introducing this chapter, I said only those who own shares in NucRocCorp and the charters should have the right to vote in the regime, and this should be proportional voting, so more shares means more votes. Some might view this as giving power to banks at the expense of the "little people." However, I have proposed a voluntary regime that has political, economic, social and other benefits, not a political alliance of nations wanting to build publicly financed outposts in Space. Private money would finance it, not tax dollars, so one man, one vote is inappropriate. Moreover, as development takes hold, investments by "little people" could outweigh those of banks and institutions, perhaps with mutual or pension funds now investing in charters. I believe but certainly cannot prove they would be quite profitable, and I cite my earlier example of the Dutch Far East India Company, which never paid less than a 12 percent dividend for almost a century.

I have discussed the roles stockholders might take in the various organizations, but here I must consider the money gathering and recordkeeping function. In the NucRocCorp I indicated the Public Accounting Office had to keep records of stock ownership, but it would be transformed later into an independent entity – a Stockholders Institution – separate from the Space Council, Space Assembly, Space Regulatory Commission, and Space Charter Authority. It would have three principal functions. First, it would keep all stockholder-invested funds in chartered entities; they in turn would have access to them as necessary to run the chartered business. So in this sense the institution would be like a bank for the chartered entity and a mutual fund for investors, distributing dividends, stock splits, and so forth. I know this will be very controversial, as corporations control their own money and just borrow from banks or issues bonds as necessary, but having a monetary check on all chartered entities will enable the regime to bring peace to the planet. Stockholders must have the right to call peoples and countries to account for actions not in accord with the membership criteria and democratic ideas, and they must have the right to develop punishments and vote on their application. Keeping their hands on the money can be a big stick in maintaining peace. I discuss this further in the next chapter. Second, its recordkeeping function allows the Institution to handle all votes (announcing their dates, candidates, issues, opening and closing the polls, tabulating and announcing the results of the vote – in sum, many things governments do for elections).

Third, since this would be a democratic space program conducted with private money, stockholders must have the right to override any Space Council, Space Assembly, or Space Charter Authority decision, the right of initiative and referendum, the right of recall of any elected official, the right to request an audit on any chartered entity, and the right to override any actions by the Space Council and Space Assembly. An important function would be to ratify the Space Assembly and Space Council decisions to open new zones. If these institutions fail to do so, the Stockholders Institution must have the right to do so. This would personally involve each stockholder in colonization, furthering the psychological sense of dominion over

the solar system. However, the details of these must be developed carefully as too much democracy, too much public involvement leads to confusion and uncertainty. The Stockholder's Institution probably should be structured like an accounting firm, keeping an eye on every dollar and detail, and not involving itself in the affairs of charters and colonization, except as mandated by stockholders.

In conclusion, my final observation is a belief or hope that after several decades a well-ingrained sense of the nuclear continuum would exist in peoples, stockholders, and officials in these four political institutions, and in governments and peoples remaining outside the regime. In other words, this would be intuitive knowledge linking development of ever more advanced nuclear engines to an exponential reduction in the costs of sending large payloads into and around Space. This knowledge in turn would further change the human psyche and its sense of dominion over the solar system, and that would have consequences on the political structures to govern their use, such as the ones I just described. Perhaps a century hence, a Space Congress would exist to govern an increasingly colonized solar system, as the millennium of human expansion off this planet continues. Those technologies and that era are not yet, so what I have proposed is useful enough to begin debate and dialogue on the infrastructures for the solid core.

# Chapter 8
## Szilard and World Peace: The Fraternal Space Regime

*This chapter analyzes the fraternal space regime that could follow from solid cores and the peace that could unfold under its auspices, one ultimately managed by citizen shareholders worldwide. It also considers how the regime could foster a nascent sense of people looking at themselves and their place and purpose on Earth differently.*

### Introduction

Achieving and maintaining peace is a subject with a long and varied history. Let me cite a few examples. We have the absolute peace of Isaiah where lambs lie down with lions. We have the tenuous peace through the rule of law by Solon the lawgiver in ancient Greece and in more modern times by Hugo Grotius, the father of international law. We have the idyllic peace of philosophers where thinkers such as Plato and Sir Thomas More create utopias where the most virtuous and wise rule. We have the illusory peace of ideologies such as Marxism were all peoples have their material needs satisfied. We have the often-realistic peace through strength best typified by Rome's thousand-year rule. We have peace through a balance of power, and a shifting of alliances to maintain that balance. We have paper peace via treaties best typified by the many arms control accords in the last half century to restrict and eliminate weapons of mass destruction. These have had but limited success over the millennia. A new possibility appeared when Leo Szilard added his seminal thought of nuclear rockets' creating world peace in 1932 and others added insight to it, such as Senator Anderson, who said they would lift men's minds from their earthbound hatreds into the solar system; James Webb, who foresaw Space allowing the development of a (presumably peaceful) universal society; and Freeman Dyson, who probably would disagree this was possible, as that is inherent in his conception of societies in conflict migrating to different parts of the solar system.

These new views have merit, but only the nuclear rocket enables them to be realized; chemical rockets certainly cannot do it, so what could emerge is a fraternal space regime, a real, not rhetorical, brotherhood of man, that gives substance to the oratory of going in peace to the Moon for all mankind by allowing all mankind to go in peace to the Moon. Yet this regime would be composed of haves and have-nots based on behavior, and allow a managed peace to emerge, or more precisely a shareholder-managed peace - not of governments, international law, or the UN – but one where citizen stockholders had the final say in maintaining it. Underneath this, a new human sense of dominion over the solar system would emerge, an embryonic social trend that reinforces the brotherhood ideal – in short, something difficult to define yet something that would appear and, I hope, lead to a universal yet unwritten agreement to place armed conflict and violence way down the list of means for resolving disputes.

## I. The Fraternal Space Regime
## A. The Contentious Beginning

As the nuclear rocket program moved toward restarting, the politics here and overseas would be hostile, much of it the old fear of "a radioactive reactor flying overhead." This would be normal and desirable, as it would sharpen the thought of those who think in Earth-Space dimensions instead of an Earth-only one. Still, this conflict would be lengthy and emotional, with every hazard imagined to be a catastrophic accident waiting to happen. Then when research and development started, every unusual, unexplained, or unexpected incident would be magnified into an accident of calamitous proportions. Here I refer the reader to the boxed text on research and development.

---

### A Note on Research and Development

Research and development is a process in which one starts with an idea but know very little about it, yet ends with a finished product that can be used reliably, safely, and economically. The beginning of research and development emphasizes the collection of data and information to form a base of knowledge, and often this phase features pushing experiments and tests as hard as possible, even to the point of having "accidents." Programs not doing this risk becoming expensive, slow, and unproductive. So at the start of this process, "accidents" are not uncommon and often are desired. However, some major unexplained incident or accident near its end is quite troubling, as it indicates something has been overlooked, underestimated, or ignored. So here criticism is not only expected, but also justified, whereas at the beginning, it can be nothing more than carping. The most famous example here is Thomas Edison and the light bulb. He had more than a thousand unsuccessful tests of filament materials, persevered and built an extensive base of knowledge, and ultimately found the right filament for the light bulb.

---

Apart from this would be deep concern from those who know what the program really meant – the chemical rocket firms and their governments who would fear jobs and a tax base soon to be eliminated. How this is handled will mark the first critical step for the United States. Only two choices exist: Go it alone (the United States and its industries) or open it up (the United States and other countries and their firms). I prefer inclusion and will let others make the case for exclusion.

As I see it, the United States must begin with two separate but related initiatives. First, the Congress must announce it is breaking the taboo (but not define what it means technically), and then send teams of experts to brief the public on what would follow (NucRocCorp, "free launches," a solar system transportation plan, the plutonium and toxic materials disposition regimes, zones and charters, a space station industry) though their exact nature must still be determined. Thus, the Democratization of Space Act would still be at the debate stage. Second, these experts, led by senior congressional leaders or NASA officials, must also be sent overseas to inform foreign governments, their space agencies, rocket firms, and people of the changes. Here, they must offer thoughts on the conditions for their participation, and give them time to debate and develop their own suggestions. When Congress begins deliberations on the Democratization of Space Act, it must invite potential overseas partners to testify, particularly those from the legislative branches. All must realize they are creating a new international regime aimed at colonizing the solar system.

## B. Membership in the Fraternal Regime

As the time to pass the Act comes, considerable momentum would exist, so the terms for membership would be critical. This might feature wrangling over conditions, with some threatening to start their own programs, but here United States leverage would be strongest, a point reinforced by the knowledge that other countries (Russia, France) had nuclear rocket programs that came to naught. Still, the simple knowledge the United States could do it could play an important part in creating programs elsewhere. So the United States must not overplay or underplay its hand.

The first question to resolve then is membership – who could join NucRocCorp and by implication the other political institutions later. Very simply, I oppose it being open to all nations, as is the case with the UN. That structure derives from the nation-state era and is open to all – "the good, the bad, and the ugly" – based on their claim of sovereignty over a body of land. Rather, I propose a moral, behavioral regime, a fraternal one, composed of the "good," excluding the "bad and ugly" unless and until they change their ways. In other words, neither a claim of sovereignty over land nor birth within the territorial borders of a country gives any claims, rights, or benefits over manmade structures built in Space or on the planets, moons, or asteroids. That exists only on Earth and does not extend into Space, except of course, if non-members do it with chemical propulsion. So in my fraternal regime I see membership supporting fundamental policy objectives and believe only democratic governments, that adhere to the rule of law, have free elections, preserve human rights, live peaceably in the community of nations, and keep in full compliance with international obligations, should be eligible to join. I expect a wide range of views to develop on this, including full, partial, and provisional memberships.

No matter what the terms and conditions end up being, they would exclude, or at least I hope so, many of the "bad and ugly" in the UN. They would become the have-nots. I find this very appealing. Countries under dictators such as Cuba, Libya, and North Korea would be ineligible. Countries under authoritarian/totalitarian rule such as China and Vietnam would be ineligible. Countries under monarchial rule such as Saudi Arabia would be ineligible. Countries in non-compliance with international treaties like the NPT would be ineligible, and this includes Iran. Countries with human rights abuses such as Rwanda and the Sudan would be ineligible. Countries with deplorable practices toward women such as those in Southeast Asia would be ineligible. However, this is more complicated than it appears, as inevitable change occurs. Some countries might see a breakdown of democracy and an upsurge of civil violence or civil war; some might suspend the rule of law; others might become aggressive toward their neighbors; others might conclude international obligations were inimical to their interests and withdraw; and still others might hover in a netherworld between authoritarianism and democracy.

That raises questions of decertifying a country, as I do not subscribe to the notion of perpetual membership. Some aggressive countries destroyed the League of Nations created after World War I to promote peace, and the same could happen here. Then there are questions of the criteria for recertification and re-admittance, including reparations to the injured party. These are formidable, and I expect extensive debate over them, but I hold they are doable. Other minimum conditions should include adherence to stronger nuclear and missile technology export control regimes, participation in the plutonium disposition regime, and conditions to prevent "fishing" expeditions, where a country joins to obtain the technology then withdraws to start its own program without any conditions.

## C. Managed Peace and Equal Economic Opportunity

I have assumed a condition for membership is that a country allows its citizens to invest in NucRocCorp and later charters. The mechanism may be the tax return form I proposed for the United States or something like the functional equivalent. This right should be linked to "free launches," so if a certain number of citizens in a country invest, it gets a "free launch." It's sort of a quota system. Thus, if a country such as Argentina joined and a certain number of its citizens, say five million, bought a single share in NucRocCorp, at $50 per share, that is $25 million; for this it should have as its reward an Argentinean "free launch" to stimulate interest in Space within its own country. *And* henceforth, Argentina would be allowed to contract with NucRocCorp's divisions for subsequent launches at cost for government purposes, and its private sector would have that right though at a price it negotiates. It also would have the right to form its own engine, payload, and space station divisions within NucRocCorp. That would push development of a countrywide space industry and build momentum and awareness; some countries and peoples moving forward, stimulating their schools and universities and benefiting economically. Yet it would highlight the growing stagnation of the countries that chose not to join for whatever reason as well as those that were excluded because they could not meet the entrance requirements. Both create voices of dissatisfaction. Join, or they must change their behavior to join, or continue to be excluded.

## The Stockholder's Declaration

Whoever invests in the NucRocCorp and subsequent Space Charter Authority should be required to sign a declaration that commits him or her to respect the purpose of the new regime, and conduct their personal lives in a manner that recognizes the rights of their fellow man. They must be made aware that failure to do so could result in forfeiture of their investment.

It is with charters, however, where managed peace and economic opportunity really develops. In applying for a charter, an entity first must organize itself, briefly stated, as preparing its corporate and management structure, developing its business plan and securing a dominant amount of funding from established institutions. After review, the Authority would issue a charter. At that point, but not sooner, I hold the public should be allowed to purchase shares in the chartered entity. The requirement that the entity first obtain a dominant amount of its financing from established institutions protects the public from irresponsible behavior. The price of shares should be low to allow the widest possible public participation even in non-industrialized countries. All shares should be held for a minimum period without redemption to allow the fledging entity to begin operations and produce a revenue stream. After that, normal business practices should be allowed, with the chartered entity listed on stock exchanges via an arrangement with the Stockholder's Institution and henceforth, bought and sold as the owner sees fit. No one, however, should be allowed to purchase shares in a chartered entity without having at least one share in NucRocCorp. I will not discuss the infrastructure required to make this work honestly, efficiently and inexpensively, but the Stockholder's Institution would have the responsibility to create one perhaps resembling the mutual fund industry where people could invest in an ever-growing number of chartered organizations. This cursory overview reveals the proposal's strength: It transform Space from a narrow, elitist program conducted mostly with public money by developed countries to a broad-based, deeply humanitarian and democratic one in which the stronger help the weaker, where those with the technological know-how help those lacking that expertise. Even the humblest could participate and profit, and the development of Space profit immensely by the creation of a worldwide, private-sector pool of funding.

Some might object, holding those in unfortunate circumstances should not spend their scant resources on outlandish schemes of dubious profitability. To that I say participation would be voluntary. Let each person determine if it makes sense but do not think for them or limit this right to only the fortunate in industrialized countries. Though they might have hard-luck circumstances, the school of hard-knocks quickly tutors its pupils, so even in the poorest of countries people could pick out the more profitable charters from those less so. Others might argue the profit potential is merely assumed and probably grossly overstated: Now, but only now, this criticism is fair, accurate, and irrefutable. However, it would change as soon as studies begin appearing, hopefully in a year or so that would show the costs of small and large engines taking payloads to LEO. Remember, for almost fifty years NASA banned using nuclear rockets to

reach LEO, and no one has ever studied the Re-core concept. So these preliminary studies would lead to more detailed ones as experts become familiar with the Re-cores and cargo plane launches, and these would give much better numbers. Then those whose criticism was once fair, accurate, and irrefutable would find their positions increasingly eroded, and if they held rigidly to them, they would find themselves dismissed from the dialogue. These newer numbers would be publicized and attract the attention of those who wish to use Space for commercial or other purposes. That would generate interest and momentum, both of which would be accelerated during hearings on the Democratization of Space Act and subsequent creation of NucRoc-Corp. Then the "free launches" would lead to the "put up or shut up" and "convincing stations," and they would lead to the creation of a large private sector space station design, construction, and operation industry. Then the situation changes irrevocably; real world experience would have been gained, and many more stations appear. Then I expect profits, and handsome ones at that.

Let me also assume another condition for membership is those countries with developed space programs must establish mechanisms to help those with rudimentary or nonexistent ones. So again, we have the fraternal nature of this new program. Some of this would have started with the "free launches," but it would be particularly important in charters, as they must be structured to prevent the appearance or fact of one nation's companies carrying out an activity in LEO and thus flying a national flag. At a minimum, a pairing of flags should exist for all charters, at least two and North-South in orientation. In other words, the chartered entity must be multinational in its corporate makeup, with management and labor drawn from industrialized and non-industrialized countries. In this way, each charter would contribute to managed peace, as each charter would include parties from different parts of the world.

This might seem unrealistic or unreasonable, given the scarce educational opportunities in some countries and the high-tech requirements of Space, but I point to the success of the U.S. military in training the raw individual, many barely out of high school, who operates an array of high-tech equipment. Then they leave and most have solid, even stellar careers in the public or private sector. I hold the same would happen here, as competent and reliable employees can be found in all countries and all walks of life, no matter how humble, and only need an opportunity for education, and once trained would serve the chartered entity well. Though multinational in composition, this chartered entity would not be a multinational corporation that often seems to operate beyond the law of any one country, evading taxes and responsibility for their actions by various dodges. No, the Authority, Space Council, and Space Assembly would have jurisdiction over it.

Let me summarize my argument thus far. I advocate a fraternal space regime that would be composed only of "good" countries that share fundamental human values and allow their citizens to invest in NucRocCorp, and the chartered entities that would be structured to pair the stronger with the weaker. Yet, this fraternal regime would confer other benefits on its members, some slanted to governments, others toward the citizen stockholder, in maintaining peace. I'll list eight to begin the discussion.

1. The right to approve charters. This has an obvious relevance to peace. As noted, I expect the charters to have a North-South orientation, so non space-faring countries could benefit immediately. Perhaps it might take the form of jobs, scholarships and training programs, then "free launch" initiatives, but as the expansion into Space took hold I expect the Authority, Assembly and Council would establish and fund regional schools and training programs, at least one on each continent, to prepare even more workers. Ultimately, the goal with charters should be a South-North orientation and then a South-South one, where countries with border conflicts or antagonisms between peoples are paired together. This North-South, South-North and South-South requirement would mitigate any criticism that the space-faring countries were imperialistic, while it would create powerful forces in less developed countries that might inhibit or eliminate egregious behavior such as tribal warfare, geno-

cide, mass expulsions, and wanton expropriation of property. When SSTOs appear, creating spaceports around the equator, their lure could be an even more powerful incentive for good behavior.

2. <u>The right to levy and spend taxes.</u> The entire point of heavy lifters and the Authority is to wean space programs off public money, creating in its place a profitable and broad-based worldwide space industry. At some point though, the chartered must pay taxes, both to their host government and to fund the regime's political structures. Obviously, the right to determine when and how they are levied and spent will be a powerful incentive for countries to join and remain members in good standing. (As I see it, this prevents international organizations such as the UN from having authority over this new regime, as they cannot tax but only receive their contributions from their member states. Some may disagree, so I'll let them make the case for giving diplomats authority to levy taxes).

Yet this will be the thorniest issue of all. For example, some of the chartered might compete with Earth-bound firms who would protest vigorously over their loss of business, even bankruptcy, but capitalism is all about rewarding the risk takers. Yet the chartered must neither be unduly privileged nor become tax-exempt. They must pay their fair share, as Earth-bound firms do. Yet taxing prematurely or unwisely might penalize the risk-takers

or kill what could become a golden goose, particularly if the steep bell curve growth I predict occurs. So an exemption must apply, but it would be contentious.

Then the distribution of taxes among the regime's member governments will become an issue. Remember I proposed that those countries that funded NucRocCorp or the first space stations must be repaid (that is, if these required government funding), so when chartered entities began making profits, any tax revenue first must go to the original funding countries. Then there's the question of tax surpluses: Should all stockholders receive rebate? Should it fund other projects, say a new zone is open and some demonstration is required there? Or should it fund schools and training programs on Earth? Or should it be distributed back to all member governments? If the latter, how? It might be as simple as one country, one share, thus making all countries equal, but that runs into the problem of unequal contributions, with the peoples of some countries investing more and their governments expecting a bigger check. So here the principle of proportionality might apply, each gaining a percentage of a tax surplus equal to their contributions, but that might favor the industrialized countries that have more to invest than the non-industrialized. To avoid this, the concept of shares used for centuries by fishing vessels might be considered, so many shares of the catch to the boat and its gear, then the captain, officers, and crew. This might be more equitable, particularly if a large share funded educational and training programs in non-industrialized countries, this equating to the boat and its gear. A wide range of views will appear here.

3. The right for public participation and remuneration. I've mentioned each country must allow its citizens to invest quickly and easily through its tax forms or other mechanisms. Here I wish to limit that right so it contributes to peace; in other words, it must not be absolute as it is in democratic countries, where property rights are enshrined in constitutions and legal systems, leaving eminent domain as the government's main tool to seize private assets. Even here, just compensation must be paid. I envision something different with managed peace where citizen stockholders, banks, venture capitalists, and lending institutions have something tangible at risk from the errant behavior of their government or from abhorrent civil actions in their country. This is akin to economic sanctions, a common diplomatic tool with uneven results, seldom changing a state's behavior but often punishing its citizens whose living standard degrades. (Officials of sanctioned governments seemingly never have their standard of living adversely affected and often profit handsomely, as the recent Iraqi sanctions demonstrated).

But, this is different, since leverage is gained over the wayward behavior of member countries or its peoples. Money could be frozen, forfeited, or awarded as reparations to the injured party; employment, scholarship, and training programs could be restricted or eliminated; tax payments to member governments could be withheld or terminated; rights to launch payloads could be cancelled; space stations could be seized; and governments could be expelled unless and until their behavior changes. That is why I said in the prior chapter the Stockholder's Institution should control the flow of investment money. By keeping its hands on the money, it could freeze assets instantly. The threat of loss can have a powerful impact on inhibiting or stopping aggressive or hostile actions by peoples and governments, and must not be discounted in its utility to maintain world peace. No doubt controversy would surround this provision from countries with democratic traditions, but that concern would be more theoretical than actual in the final analysis. These governments have established safeguards against abhorrent civil actions and normally begin military operations reluctantly, after prolonged debate, and in accord with international obligations.

Let's take an example and assume Russia was a member, and it and its citizens had invested in "free launches" and chartered space stations. If Russia were to invade another country as it did Georgia, without provocation and without trying to settle the dispute in the UN or other diplomatic or international forum, the Stockholder's Institution could freeze all its assets in the space regime rather quickly. (The UN provides procedures for resolution of disputes, other than military force. After the UN proper, there are direct negotiations between the parties under established rules of diplomacy, the intervention of a disinterested, neutral third party, and referral to the International Court of Justice. None are binding on the disputants however). It and its citizens could not obtain their money and eventually might have to forfeit some or all assets – including space stations - as reparations and perhaps be expelled from the regime. This ever-present threat of loss might have had a deterrent effect on Russia, but if it didn't and its people suffered a loss of assets, they would probably press for immediate changes in the government. Keeping good behavior might not be a powerful tool at the regime's start, but if the bell curve growth I predict occurred, it could be significant.

4. The right to admit and expel members and determine punishments. I discussed this previously in the context of member governments developing membership criteria and punishments. Yet a more powerful dimension exists whereby citizen stockholders could have their say on admitting and expelling members and determining punishments. This would be fundamentally new in international politics, and it's very simple to envision. Since stockholders would have voting rights proportional to their shares, I expect groups would organize to promote a particular cause or goal, sort of like the Non Governmental Organizations (NGO) in the UN. But this could be different. They have votes and if they gain enough of them, they could exercise authority over three of the four political institutions (not the Regulatory Commission). Using my example of Russia, if the political institutions failed to act, these Space NGOs could band together to expel, punish, or order reparations, and make the institutions enforce it. Probably the more astute among the peace and human rights and environmental movements would seize this opportunity immediately, but it would spread to other groups, such as those concerned with taxes. How to organize this stockholder authority is debatable and how effective it could become is uncertain and depends on how it is structured. But NGOs have no voting rights in the UN, and the UN itself has limited authority, so anything is an improvement.

5. The right to regulate. This might not be a strong incentive to prevent wayward behavior by states or peoples, but it could contribute to a state and people keeping their good behavior if its citizens have a role in the Regulatory Commission.

6. The right to adjudicate. As commercial development increases, with ever more people going into Space, commercial disputes and criminal behavior are inevitable. Either the existing international legal system would be adapted to the new political institutions to deal with them or the Space Council or Space Assembly might have to create them. This might have a role in managing a peace

depending on how it is structured. My sense is the Space Council and Assembly would create a legal system to have authority over it after gaining experience with the "put up or shut up" and "convincing stations," and as the outline of the bell curve growth in space stations becomes clearer.

7. <u>The right to open up new zones for development.</u> This would have long-term importance and be a powerful incentive to remain a member in good standing: As progress is made along the nuclear continuum so are new zones opened up. For example, today many view the Moon as an economic powerhouse, and if opened for colonization, those excluded countries would grow increasingly unhappy as they literally could see economic benefits flowing to others while they suffer stagnant economies, uncompetitive and obsolete industries, and the migration of their educated elite to regime member countries. At some point, that dissatisfaction would reach a critical point, and the state petition to join the regime; then ever more stringent conditions for membership could be imposed, ones that contribute to managed peace, such as giving up nuclear weapons programs. Also, as other zones are opened, it means a reduction in the costs of reaching ones already open, spurring additional development and profit there, thus furthering the unhappiness of those excluded, and reminding those who are members they have a lot to lose by wayward behavior. Linked with this would be

the subtle point of a psychological transformation I've mentioned that will occur in peoples everywhere, beginning with the solar system transportation system and continuing with the various space stations. A newfound sense of ownership and dominion over the solar system, a sense that gives rights to the individual and countries with good behavior but none for those with bad.

8. <u>The right to use the space stations.</u> I've discussed this before, but let me add a twist here. The "put up or shut up" and "convincing stations" could have 25-50 year life expectancies, but would have served their demonstration purpose in 5-10 years. This raises the question of selling them or allowing the members to use them. If the latter, in troubled areas, peace initiatives might be advanced by sending representatives of disputing states or peoples to either station to try to resolve their difference while gazing down on the planet. This could be done at the request of international organizations, such as the UN, by countries in dispute, or by a country being torn apart by civil strife. It might seem farfetched, but the "Earth-rise" picture taken by Apollo 8 has had a long and powerful impact on people on Earth, particularly the environmental movement. So it might not be so farfetched after all. Sometimes the location of talks produces significant results, and this is a reason why peace conferences are often held in neutral countries.

**EARTH RISE**

CREDIT: NASA

The Apollo 8 photograph of the Earth rising over the Moon's surface has had a profound effect socially, particularly on the environmental movement and the awareness of the fragility of the planet. The same effect might be used for peace by sending the parties in a hostile dispute to a space station where a new perspective might lead to solutions hitherto not possible on Earth.

Another option might be a raffle in which all member states vie for the right to use the stations for a time at cost. How they choose to use their turn would be their prerogative. Some might use it for national purposes such as science and education; some might sell the right to the highest bidder; some might conduct a lottery, pocketing the proceeds above the costs; some might trade up or down on the queue for better terms; some might form partnerships with other members. Such looming profits might give coun-

tries and peoples pause before embarking on wayward actions.

These topics must be developed, and I hope my overview will stimulate others to do it. There is something profound about nuclear rockets and their consequences, and their viewpoints and expertise would add to the debate and dialogue. In doing this, I hope I have given some scope and depth to Szilard's thought of nuclear rockets saving mankind from perpetual war. Indeed, one may wonder if NASA hadn't established the ban whether the killing fields the world has seen since 1960 could have been prevented or mitigated. I don't know. Had it not been established, I believe the space program would be quite different. Many space stations might be flying in zone 1. Perhaps Szilard might also agree with the four political structures I've set forth, with their global system of interlocking vested interests and punishments and haves and have-nots. I think he would, I think he would like this "carrots and sticks" regime, or at least I hope so.

## II. Nascent Fraternal or Social Trend

The preceding discussed how a peace could be managed but a fraternal or social dimension will develop, influenced by the investment opportunities and sense of ownership over the solar system, by the media who spend time on the "convincing station" and then help formulate public opinion and by writers, some of whom might become the new Homer or Virgil as they tell their vision of this new human endeavor. Then average citizens would go there, buying tickets or winning them in lotteries. This unleashes a dynamic, one that I think would give people everywhere a more benevolent perspective on themselves, their neighbors, and their beliefs and actions. Perhaps this is what Senator Anderson had in mind when he said nuclear rockets would lift men's minds from their Earth-bound hatreds into the universe, implying as the mind's gaze shifts upward a rethinking occurs that looks ever more at what unites us as humans than what divides us as peoples. In a sense, this is like sailors being equal in a hostile sea, giving rise to the requirement that those at sea must aid those in danger no matter the color of their skin, their religious beliefs, or political affiliations. I find this new, emerging dynamic hard to discuss, as it is so subjective, but I note it will happen, as all technologies have social and intellectual consequences. I can only offer a few thoughts on it, how it could grow, how it could be fostered, and where it might lead.

### Anderson Lights

In honor of Senator Clinton P. Anderson, all space stations should be required to have lights visible from Earth clearly identifying the space station, blinking out its ownership or messages in Morse code (also a good source of advertising revenue). Having people all over the world gaze upward and see them circling overhead day and night would promote investment opportunities and reinforce the democratic ideas of equality and opportunity. All could participate by owning stock, and by traveling to the stations.

Scientists have had the reputation over the centuries as some of the most eloquent spokesmen for peace, in addition to their roles in shaping the intellectual life of their countries. That role could be dramatically enhanced after the "put up or shut up" and "convincing stations" by constructing a Super Size Science Space Station. I know it's a dreadful name and again I show my inability to coin useful ones, as this sounds like a toy or item on a fast food chain menu, but perhaps it could be called after some of the scientific greats of the twentieth-century: Einstein, Rutherford, Bohr, or Szilard. This station should be quite large, perhaps 10,000 to 20,000-tons, enough to house hundreds for months at a time, and contain all the scientific and engineering disciplines a large university would have with all their scientific equipment. It would have many functions: To conduct experiments, obviously; to beam back to Earth daily reports on the weather and warnings of solar flares; to construct scientific payloads on-board and launch them into the solar system quickly and easily; to create an asteroid and comet warning system around Earth and if necessary to function as a command center to deflect or defeat those on a collision course with Earth; to be a classroom in the sky, an institution of higher learning (pun intended), beaming back courses to participating universities; to offer training for talented undergraduate and graduate students and opportunities for professors on sabbaticals to spend some time in residence there. I suspect today's senior scientists would not look with favor on this, as they would be retired when this station begins operations perhaps a quarter of a century hence. Thus, when NucRoc-Corp was formed, it and NASA's Offices of Space Science and Space Stations with their counterparts overseas must form study groups with younger scientists in their twenties or thirties to work on the design of this super station. When it appeared, they would still be young enough to spend some time in their creation. All of this would be government-sponsored and funded, though the private sector certainly could have a role, such as endowing chairs of higher learning (again pun intended). If built, it would allow scientists to conduct research of unquestionable value to the world, and it would give them a collective voice of inestimable worth in commenting on the peace issues of the day. Perhaps that should be a requirement, to give an annual state of the peace speech, which may be of even more value than the scientific research.

Next, a consequence of chemical rocket space programs is that most people are passive while the elite do things for them, giving pictures from the Hubble telescope, stunning thoughts on the creation of the solar system and universe, and heroic astronauts basking in ticker-tape parades. In contrast, nuclear rockets empower the individual, a change in attitude the "free launch" and charters would accelerate. Now they can have a piece of the Space action, from the most inexperienced to the most educated; in other words, everyone could be a Space capitalist. So one consequence of nuclear rocket's allowing dominion over the solar system would be a newfound sense of the individual's right of ownership over it, the right to participate in and profit by its development, the right to go there. Rights are powerful things in the political, economic and social life of a country since our founding fathers gave political birth to them over two centuries ago, and they have spread throughout the world. Where this newfound sense of rights would lead is uncertain; perhaps some twenty-first century Thomas Jefferson might pen a document like our Declaration of Independence that eloquently extends our individual rights to the end of the solar system. That could have a galvanizing effect, as the Declaration has had.

Yet another consequence could be a newfound sense of the purpose of man's existence, perhaps resurrecting it around Konstantin Tsiolkovsky's famous comment that Earth was the cradle of civilization, but man does not live in the cradle forever. Or it might center on Robert Goddard's thought of humankind escaping into the universe from a solar system catastrophe. Or it might just include a renaissance of teleology, the proposition that the universe and man in it have design and purpose, to challenge the randomness of Darwinism and natural selection. These can be extraordinarily powerful, much like the democratic ideas of thinkers in the seventeenth and eighteenth centuries led to the overthrow of kings and queens, czars and emperors. This would take time to develop and permeate through societies, but it would, and when it did it could easily turn into resentment or stronger actions if individuals perceived inequitable or unfair treatment or if the

Authority, Space Council and Space Assembly were unresponsive to their requests.

This suggests the new political institutions must aggressively obtain the views of the citizen stockholders over and above those initially designed to promote the development of Space. With the Internet now available worldwide, one mechanism might be polling, with the results made known in all languages. Another might be holding frequent hearings in the capitals and cities of member countries to identify what stockholders want in the future, an activity good for marketing as well. These may mean a dynamic leading to a different global society was underway, as people everywhere begin to think and act in Earth-Space dimensions, so this might be James Webb's idea of a universal society. And as they integrated this new capability into their lives, they might see opportunities others might not. Rather than being denied a charter, they should be given one, provided they find the financing and inform potential investors of the risk. Here a manned Mars mission might make sense, as a deep and experienced pool of management, labor, and capital would be available to make it happen. Another might be a group wanting to migrate to different areas of the solar system to follow their beliefs (though it's still hard to see Freeman Dyson's idea as practical unless fusion engines become available). Yet another variation might be a religious order wanting to establish a monastery in LEO. Here, the North-South, South-North or South-South orientation requirement should be waived.

## William James, Leo Szilard, and the Moral Equivalent of War

In 1906, the American philosopher William James delivered an unusual commencement address at Stanford University, denouncing war as unthinkable in the Machine/Industrial age, but praising its virtues, such as heroism, duty, honor, and self-sacrifice as vital to the well being of a nation. He rejected pacifism as harmful to a nation's health unless it could provide an alternative way, a moral equivalent, of producing such virtues. His solution was national conscription, drafting the young to work in any number of hard, manual labor jobs for a time "to get the childishness knocked out of them, and to come back into society with healthier sympathies and soberer ideas." This moral equivalency idea has had a powerful impact on American life, notably leading President Kennedy to establish the Peace Corps in 1961.

I do not know whether Szilard read this essay or even heard of the moral equivalency idea, as he lived in Europe when he said in 1932 that man would war forever on Earth unless provided an outlet in Space. He was only 34 then. He mentions neither national conscription nor essential national virtues, yet his comment contains the essence of moral equivalency, but with a twist. He would substitute a radically new, good activity for a millennia's old, bad one, whereas James merely substituted hard labor, man's lot for eons, for war. I believe it is quite plausible that using nuclear rockets to leave Earth permanently would involve virtues such as heroism and self-sacrifice. Danger abounds in Space in the form of sudden death.

What I propose lies within this substitution tradition, yet it is different. Governments would start the ball rolling by forming NucRocCorp and the four institutions, but once rolling, stockholders would exercise control over the creations via the democratic process. (With James, this was always a government responsibility via national conscription for hard labor). In due course, they would have the authority to exclude the "bad and ugly," and punish errant behavior by peoples or governments in that regime. So sovereignty would fall to them ultimately, and they would be charged with having everyone respect fundamental human rights, which is much broader than preventing war.

Would this new fraternal space regime lead to a total and complete cessation of violence and war on Earth, a peace of Isaiah, if that is what Szilard had in mind; would it be a harbinger of Webb's universal and presumably peaceful society? My answer to the first question is probably not, I regretfully must say, as man is unique in his ability to find ways for violent conflict. What it could do is limit those options dramatically, so whatever violence did arise between its members would be shorter, less vicious, and more easily challenged by stockholders. If so, this could well become Webb's universal society, but tempered quite a bit. It would be imperfect, yet the prospect of losing material possessions can be a powerful incentive for behavioral change – witness the New Testament story of the woman who dropped everything to find a lost coin. For those outside the regime, the pressure would increase dramatically to change their behavior, or continue to lose while the rest of the world prospers. Yet in the end, this fraternal regime could only produce a managed peace, not one where lambs and lions lie down in perfect harmony forever, but where stockholders have the final say in preventing or mitigating war and conflict. That, indeed, would be epoch-changing.

# Chapter 9
## Summary and the Way Forward to the Legislature

*I have put forth ideas to liberalize the space program, to extend its dominion to the end of the solar system, to make it progressive by harmonizing it with the ever-increasing power and speed of engines on the nuclear continuum, and inclusive by allowing individual citizens a right to participate in and profit from them – directly by flights into Space or indirectly by investing in the NucRocCorp and chartered entities - to change its funding from public to private money, thus creating a worldwide, broad-based space industry that governments can in turn tax, and to structure it to lead to world peace by eliminating plutonium and by excluding countries harboring undemocratic values or aggressive tendencies, and by lifting men's minds from their Earthbound hatreds into the heavens. In sum, the space program I propose is inexpensive for governments and egalitarian and equitable to peoples; in other words, it is democratic in its conduct with dominion as its goal and introduces a new epoch in world history. With this, I think I have covered in a logical and rational way the themes and ideas I listed in chapter 1. I hope all see by now that breaking the taboo could have profound consequences.*

### Comparison of Chemical and Nuclear Rocket Engine Space Programs

| Chemical | Nuclear |
|---|---|
| * Elitist: participation forever limited to select few for manned and unmanned flights. | * Egalitarian: space program increasingly democratic, opening up participation to more and more. |
| * Expensive: high cost of reaching LEO requires government funding or subsidy or capital rich private sector or small payloads or combinations thereof. | * Inexpensive: increasingly lower costs of reaching and moving beyond LEO stimulates private sector involvement while reducing government funding and changing its role. |
| * Inequitable: only a few benefit from space spending, little direct benefit to taxpayers from government spending. | * Equitable: all taxpayers can benefit from space spending directly. |
| * Exploration: space program forever limited to exploring solar system. | * Colonization: space program increasingly facilitates colonization of solar system. |
| * Public funding of space program forever: results in shrinking space program with loss of jobs and tax base. | * Private funding of space program: creates many new jobs and broad tax base in advanced and developing countries. |
| * Executive branch paramount: with limited policy options to solve earthly problems via one-dimensional thinking. | * Legislative branch paramount: ever wider array of policy options to solve earthly problems via two-dimensional thinking. |
| * Plutonium and toxic wastes: remain on Earth, perhaps forever | * Plutonium and toxic wastes: disposed of permanently in Space. |

In so doing, I categorically reject the static "mission-itis" thinking inherent in single-shot chemically propelled rockets, replacing that with thoughts of an integrated solar system transportation plan using small Re-core and Re-use engines, thus beginning man's dominion of the solar system, and of a zones concept for larger ones, thus extending it as permanent human settlements progressively appear there. Also, I categorically reject all commissions, reports, and studies on space policy, including its many "visions," as inherently flawed as all ignore heat and weight. Specific impulse really does matter after all; the gravity well is deep; and the rocket equations are real. Don't ignore them any longer: Stop gold plating the nail; get a bigger hammer, and put the ponies out to pasture. Finally, I categorically reject all self-appointed protectors of the Earth who say nuclear rockets must never be used to reach and return from LEO. I say stop disparaging the scientific, engineering and technical skills of our men and women who can do it. I say protect our planet by moving off it, by creating new industries in Space, by helping solve global warming through Space-based solutions and by allowing expansion of the nuclear option here on Earth, by shipping strategically dangerous and environmentally harmful materials off it, and by giving jobs, education, hope, and human rights to all, particularly the poor and vulnerable in the world. Think in two dimensions, Earth and Space, not one. Don't keep our planet dirty, warming and warring. Give peace a chance, and let citizen shareholders ultimately manage that peace. Power to the people!

---

### To Break or Not to Break the Taboo

In plain English, the choice is simple. To not break the taboo means to continue in "mission-itis" thinking - one-shot, one-mission - and to justify the nuclear rocket program on the basis of that mission, which for too long has been manned Mars. That hasn't worked in half a century. To break the taboo means to justify the program on its many consequences: Diplomatic, political, nonproliferation and disarmament, scientific, educational, energy, environmental, technological innovation, private investment, industries, jobs, a broader tax base, and social – the creation of a fraternal space regime that features managed peace among the "good" while excluding the "bad and ugly" unless and until they change their ways. I hold the public will decide such benefits far outweigh any potential risks, and support the program.

In putting forth thoughts for a fraternal space program, one hallmark of democracies is debate and dialogue. So I say look at my ideas not as rigid and inflexible, but as offering a new way of thinking, aimed at creating infrastructures first and foremost, and doing so in a controlled manner over several decades, not in a rush to a planet, or an outpost on the Moon, letting infrastructures adjust to and digest the new capabilities. Here I recognize technology has political, social, economic, intellectual, and religious consequences that dialogue will reveal further. Using bureaucratic terms then, think of this book as putting a marker on the table, to provoke debate and discussion. So agree, disagree, change, modify, propose, adapt, adjust, amend, then realize this discussion could never be happening with chemically propelled rockets or with so-called "advanced" ideas like space elevators, yet most of all listen, and not just to one another, but to the public who might enter the debate slowly, but who are its ultimate benefactors. So run the numbers with Re-core/Re-use engines and cargo plane launches and heavy lifters, develop competing solar system transportation plans and zone concepts, discuss different corporation structures, membership criteria and methods for allowing taxpayers to participate. Then broaden the discussion to include those who would benefit from the "free launches," then those who want to build giant space stations and those concerned with plutonium inventories, and hazardous and toxic materials. Then go to the public.

This discussion must begin in industry, government laboratories, and academia, perhaps fostered by NASA, where people have expertise, resources, and time to consider a new way of thinking. Form a group to direct the studies, and host conferences to discuss them. Then publish white papers for the public, and spread the message via outreach efforts to professional organizations and societies, letting those with business, management, taxation and other expertise provide their views. In this process, I hold that an overall plan will crystallize and a group of bright, young, articulate leaders, a society of breakers, emerge to foster the debate and engage the public via television, newspapers, magazines, and other forums and roundtables where people exchange views. To these who break the taboo I say do not be afraid, do not say or even think we cannot do this, as that is not realist but defeatist thinking. I hold the public is neutral, but will embrace its sons and daughters who think in a different way about Space. They just need to hear the rationale. So go to Oprah, engage the public, let them see the better argument: Those who fear a "radioactive reactor flying overhead" or those who seek a democratic space program, one bottom up and not top down, where all could participate and benefit and a managed peace could emerge. Let them judge. At some point, this leads to Congress and not NASA, and this group of breakers must bring it there. If the goal is to create a democratic space program, the legislature must begin it. It cannot start any other way.

Archimedes, the most famous scientist of antiquity, once said if he had a place to stand he could move the world. Of course, he thought the Earth was stationary and with a platform outside it, he could use his levers and pulleys and move this large motionless mass. We now know the Earth rotates on its axis as it moves around the Sun, making his belief impossible to achieve. Yet engines on the nuclear continuum give his thinking renewed life and vitality, not so much for scientists, but for legislators who will become Archimedes's successors, moving the world in ways he never could have imagined, ushering in a millennium of human expansion that brings with it perhaps not a universal society, but certainly a fundamentally different and more fraternal one. Thus, those breakers who lead this discussion must request Congress to create a permanent forum for debate and dialogue. That is crucial.

As the legislators listen and question, the breakers must impress upon them the awareness they will not be just men and women who pass laws, authorize and appropriate money, and oversee the executive branch in the traditional nation-state structure; now they, with legislators of like-minded countries, can become architects of a new epoch-changing era, one that controls and directs man's permanent entrance into the solar system. If done properly, it could help maintain peace and stability among nations.

## The New T-Shirts

Once Congress breaks the taboo, the legislature becomes the driver of the space program; it becomes the architect of a new world order, not presidents, secretaries of state, NASA administrators and especially, not authoritarian leaders, dictators or so-called liberators. Duly elected lawmakers here and in like-minded legislatures overseas will become the new revolutionaries of the third millennium. T-shirts might soon appear, saying "Che is dead. Democracy lives and it is going into Space." So the blight of communism will finally disappear, as the appeal of a new epic for humankind takes root.

Calling this new regime imperialistic is nonsense. Empires have a ruler and the ruled, though subjugated is a better word, as coercion is always present. They have no say in their government or freedom to leave. Yet this fraternal regime would be voluntary – no nations or peoples would be forced to join, and anyone can cash out at any time. Yet in a reverse way, coercion would exist – countries or peoples could be thrown out and their money confiscated for violent, non-peaceful behavior on a vote, ultimately, by their peers.

Breaking the taboo, then, might lead to the peace of threes: The peace of a millennium of expansion into the solar system, occupying the minds and energies of peoples everywhere; the peace of a millennium of prosperity where all can reap the rewards of building new "Earths" throughout the solar system; and the peace of a millennium of fraternity where the strong help the weak and the educated those less so, creating a voluntary brotherhood of disparate peoples in the world. It would be the American experience of E Pluribus Unum now writ large, but with legislators from different countries doing the writing.

I hope by now many have been persuaded to pursue this new way of thinking, but others might still be scoffing, saying this must be rejected as a simple monist scheme, reducing everything to a single factor, and therefore is out of touch with our complex world. To them I say history teaches otherwise. It's happened before. A thousand years ago the Vikings' mastery of the long ship, a single factor and the high-tech machine of the time, gave them dominion over the foreboding oceans, though they used it to pummel Europe for centuries. Our mastery of nuclear rocketry, a single factor, will give us dominion over a forbidding solar system, allowing human expansion there; if used wisely, it could spread democracy across our planet. Others might say this was all an illusion, a fantasy, not a vision, and any lights seen whirling overhead in the sky would more likely be UFOs than space stations. I urge them to think again. The rocket equations and specific impulse are not figments of anyone's imagination, but measures of technical reality. Increasing temperature while reducing weight lowers the economics of going into Space. It's as simple as that. Break the taboo, and this human expansion happens. It's inherent in the technology. When it does, the space program invariably changes, and somewhere on the nuclear continuum, massive space stations housing millions become feasible. Those who say that could never happen will inherit Moonshine, the one word epithet the most famous atomic scientist of the first third of the twentieth-century hurled at the young Szilard in the 1930s when he foresaw the industrial uses of atomic energy.

**SPACE STATIONS**
CREDIT: NASA
In 1929, John Desmond Bernal proposed a hollow sphere 16-kilometers in diameter that could hold 20,000-30,000 people. In the mid-1970s, Princeton physicist Gerard K. O'Neill developed the "Island 3" space colony concept. Composed of a pair of cylinders, each 20-miles long and 4-miles in diameter, giving a total land area of 500-square miles, it could house several million people. Both are impossible to build with chemical rockets, and quite expensive with solid cores, but somewhere along the nuclear continuum are engines that make such structures feasible.

If they persist in ignoring technical reality, then I say "Stand aside. Be silent, spend no more time on things that can't work." Let others who share this new thinking make it happen. So in the end I stand with Szilard, who glimpsed a greater reality for human society; I stand with Ulam, who saw something more revolutionary here than the H-bomb. I stand uneasy, though, because a more fundamental question emerges out of this: Why do it at all? Discussing that requires another book, and it involves subjects different from specific impulse, nuclear physics, Re-core and Re-use engines, political structures, charters, and voting. I speak of the stunningly swift destruction of the sixteenth-century world view - a seven day creation of a Ptolemaic world that stood for some 1400 years, with Earth as its center, covered by a heavenly vault, with hell lurking somewhere beneath – to one where Earth is just a planet in a solar system in a galaxy in an expanding universe full of galaxies. Why do it at all then involves discussing man's role in this new reality, where he can have, within several decades, nuclear engines powerful enough to leave permanently, beginning a millennium of expansion into the solar system. And after that, when even more powerful nuclear engines appear, comes star flight.

**PTOLEMAIC WORLD (LEFT) AND MILKY WAY GALAXY (RIGHT)**
CREDITS FOR BOTH: NASA
In 1539, the German astronomer Peter Apian produced his Cosmographia that depicted the Ptolemaic view of the heavens, with Earth at the center surrounded by the planets, Sun and stars. Already under assault by astronomers such as Nicolas of Cusa, this worldview was completely replaced within a century by Copernicus, Galileo and Kepler. Now our worldview is completely different, as illustrated by a NASA's artist version of our Milky Way, with our solar system located near the rear of one of its spiral arms.

# Appendix A
## Technical Development of the B-4 Core: The 1960s

### I. Introduction

The reader must know the technical basis for my argument before accepting it, so I offer three appendices to that end. This one reviews the B-4 core and its test history. For more data, I suggest reading To the End of the Solar System, as it discusses in greater detail Rover/NERVA's technical history. Appendix B considers a likely development path of first-through-fourth generation engines to justify the payload numbers I use. The reader must see that they weren't pulled out of the air, but have a basis in technical reality. Appendix C is a ten-step walk-through of the flight profile to LEO, considers the hazards in each step, and discusses how they could be overcome or mitigated.

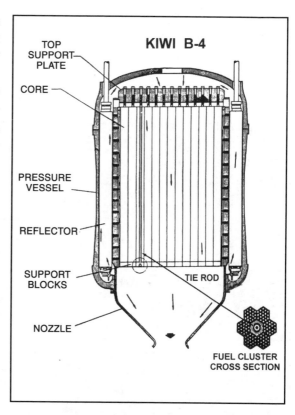

KIWI B-4

TOP SUPPORT PLATE

CORE

PRESSURE VESSEL

REFLECTOR

SUPPORT BLOCKS

TIE ROD

NOZZLE

FUEL CLUSTER CROSS SECTION

### II. The B-4 Core

The B-4 featured the hexagonal fuel element, three-quarters of an inch from flat-to-flat, with 19-propellant holes in it, each 0.10-inches in diameter, and 52-inches long. These were grouped in clusters of seven, one surrounded by six; the center one had no uranium but a tie rod or tie tube that ran the length of the element, with nuts on both end to screw the other six fuel elements into a tight bundle. Bundles were stacked together to comprise a core that resembled a beehive when

**B4 CORE**
CREDIT: NASA
In 1959, skeptics feared the tie rods holding the fuel bundle or cluster together (lower right) could break, allowing the core to be ejected out the nozzle. However, the B-4 design soon won them over, and in 1961, when Westinghouse entered the program, it embraced the design wholeheartedly because it could be "ruggedized" for flight. In its first test, the KIWI-B4A ejected a third of its core, but that was corrected, and in all subsequent tests the B-4 core proved its soundness.

viewed from above. The largest core was 55-inches in diameter and the smallest 13-inches. The most common size was 35-inches, with one tested at 20-inches and a short, stubby one on the drawing boards to be 37x37-inches.

## III. Sample of B-4 Tests

The KIWI-B4A disastrously eject-ed a third of its core out the nozzle in November 1962, but that was correct-ed, and KIWI-B4E proved its stability in a full power run in September 1964. Then it restarted for another full power run, which was a first and promised a revolution in mission planning, as it meant the engine could be restarted for mid-course corrections and maneuver-ing. Aerojet and Westinghouse "ruggedized" the B-4 for flight, an effort that paid dividends when its NRX-A3 reactor momentarily ran out of LH2 in 1965 and temperature sen-sors raced off the scale, implying the

**THE PLANNED ACCIDENT WITH THE REACTOR IN UPPER LEFT, EXPLOSION IN UPPER RIGHT AND BLAST AREA ON BOTTOM**
CREDIT: LOS ALAMOS NATIONAL LABORATORY

Flight safety was always a concern, particularly a reactor "blowing up" on the launch pad or in flight. To aid safety research, Los Alamos rigged a 1000 MW reactor to have a super fast run-away chain reaction to blow it up. This involved reworking the con-trol drums to rotate 100 times faster than normal, and was no easy feat. In January 1965, and in 0.8-seconds, KIWI-TNT "blew up" with a force estimated to be the equivalent of 300-pounds of black gunpowder, not high explosives. Damage was minimal, a broken window in a trailer several hundred feet away. Cleanup crews waited several days to let the radioactivi-ty decay to lower levels, then picked up the pieces with tongs. KIWI-TNT was, in effect, the twentieth-century version of the nineteenth-century steam boiler explosion. Never could it have been a nuclear weapon explosion. Those occur in shakes or millionths of a second; TNT was just too slow at 0.8 seconds.

core might have overheated and been damaged. It wasn't, and A3 was tested again without trouble. All subsequent tests used the B-4 design, and no core problems were encountered with it.

Other notable successes include the NRX-A6, running for more than an hour at full power without any problem, and NRX-EST (1966) and XE-Prime (1969), transitioning from reactor to engine tests. The latter two wrote several chapters of an engine manual to tell operations staff exactly what to expect under various mission scenarios, with full power, maneuvering, stops and starts, and idling capabilities. The next steps, never taken, featured tests to integrate the engine into a stage (engine plus its propellant tank) for flight-testing, the last step before being declared operational. There were no insuperable problems here. The NERVA I engine was to have 0.997 reliability, which doesn't mean it would have a catastrophic accident three times out of a thousand, but only that it failed to perform as designed. Indeed, the KIWI-B1B test in 1962 is instructive here: Though its core design was initially the favorite, some partial tests of it showed its instabilities. In a full-scale hot test in September 1962, KIWI-B1B ejected fuel elements as it started up, but the test run continued all the way to full power

with B1B, still ejecting fuel, then all the way back to shutoff, still ejecting fuel. Control was never lost - the reactor kept churning out power even while ripping itself apart. This holds out hope of completing the mission or saving the astronauts if an accident were to occur.

## III. B-4 Development Potential

In the text, I discussed two policy-oriented reasons out of five for picking the B-4 core. Here are the other three. The third is that the B-4 core can be made extremely reliable – no one doubted this – and this reliability cannot be guaranteed for other solid core concepts. Indeed, if a major accident occurred, the original NERVA I featured emergency mode operation to permit an engine thrust of 30,000-pounds at 500-seconds of specific impulse. Fourth, it has the flexibility to be scaled up or down according to the power level desired, the largest being 5000 MW, the most common 1000-1500 MW, with one at 500 MW, and the smallest being used only to test fuels. Other concepts may lack this flexibility, and so what may seem better initially may not be so subsequently. This flexibility may become even important when heavy lifters are developed; some might become super duty heavy lifters.

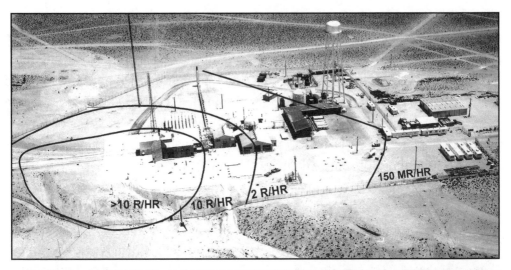

### THE UNPLANNED ACCIDENT.

Credit: Los Alamos National Laboratory

In June 1965, several months after KIWI-TNT, a real accident occurred when the Phoebus-1A ran out of LH2 while at full power, overheated, and ejected part of its core out the nozzle. Los Alamos cordoned off the area, made a radiation map, then waited six weeks for the radioactivity to decay to more manageable levels. Then cleanup began. It took 400 people two months to complete at a cost of $50,000. Their exposures would be under the limits for today's nuclear industry workers.

The fifth reason is the B-4's growth potential, and here I do not mean just building a bigger engine, but getting more power out of an existing one. This is power density, something that chemical rocket engines do not have. It's the power or thrust derived from a given volume, and is seemingly ignored by today's nuclear engine designers. For example, a first generation NERVA I would have had a power of 1500 MW or 75,000-pounds of thrust, produced from a B-4 core 35-inches wide and 52-inches long. Yet even then, everyone knew a second generation NERVA I would reach 2500 MW or 125,000-pounds of thrust, still in the same core dimensions, still weighing 15,000-pounds. At one time, Los Alamos thought the 35x52-inch B-4 core could produce 4000 MW, or 200,000-pounds of thrust. On the face of it, this is merging or clustering two or more engines into one, ostensibly without increasing its weight.

---

**CLUSTERING**

In the early days, many felt nuclear engines could not be clustered together like chemical rocket engines to give greater overall thrust for the mission. Many feared neutrons from one reactor would penetrate the other, cause unwanted fissioning, and jeopardize control of the engines. Los Alamos proved this fear groundless in a series of experiments in 1964 and henceforth, many mission studies featured nuclear rocket engines clustered together.

---

This is impressive, but must be qualified, as power density is a subtle topic, and not as easy as it sounds. In general, it can occur by going to an all-loaded core, that is, replacing the center unloaded (no uranium) fuel element in the bundle with a loaded one. That increases power by one-seventh, to 1870 MW. Or it can occur by enlarging the propellant holes or by having more propellant holes to allow more hydrogen to flow through the core. Or it can occur by operating at higher pressure, that is, pumping more LH2 into the core. Now the subtle problems arise. Going to the all-loaded core means developing a means to keep the fuel

bundle and core intact; it implies scrapping the metallic rod or tube that hold the bundle together for some other scheme. Increasing the size of the propellant hole or adding more of them means potentially weakening the thickness of the web between the holes, perhaps reducing reliability. Pumping more LH2 into the core means putting increased stresses and strains on the entire engine. And increases in power density mean increases in thermal heat, radiation, and other stresses and strains, so the entire engine must be strengthened. All this can add weight, which detracts from the power increase, so it's not as simple and straightforward as it sounds. These are engineering matters, however, not insuperable barriers.

This discussion might seem confusing to the general reader, so I'll illustrate power density with an example from my hot-rod days, which I think most would understand. At the age of 18 and finally free from my father's clutches, I bought a 1949 Mercury convertible with a flathead V-8, rated at 110-horsepower with a two-barrel carburetor. I added another two-barrel carburetor, bringing its horsepower up to a whopping 120. Painted candy apple red with a white top, it was hot and made that sweet rumble only a flathead can make – similar to a Harley Davidson motorcycle. I still lament I sold it when I started college. Ford stopped making flatheads in 1953, but hot rodders never stopped working on them. At a bookstore recently, I saw a magazine featuring flatheads with over 500-horsepower. So here then is the flathead's history: Ford introduced it in the 1930s at 60-horsepower and up-rated it over the years to 100-horsepower for Fords and 110 for Mercury and Lincoln before terminating its production in 1953. Yet hot rodders continued to work on them, now getting over 500 horsepower. Yet it's the same size engine block. Here then is the similarity to power density. By concentrating on the B-4, engineers will develop that expertise that senses or knows intuitively what can be done and how to go about doing it. It is dead wrong to assume the B-4 will reach its technical limits after several generations, with no other improvements possible. Perhaps an eighth-generation system might have power densities larger than 4000 MW yet still in the 35x52-inch core.

Compared to power density, fuel element development is equally important, and has two parts. One is the ability to increase in specific impulse; the first NERVA I was to have had 825-seconds of specific impulse and a second generation, 925-seconds. The latter would mean operating around 2400-2700° C. With additional development of these graphite fuels, a 1000-plus seconds is possible or operating above 3000° C. The other important part of fuel development is its lifetime. The NERVA I would have had 10-hours of full power operation with 60-recycles or stops and starts. Both are tremendously important for missions. I continue this discussion in Appendix B.

# Appendix B
## Development of First-Through-Fourth-Generation Engines
Stan Gunn (Rocketdyne), John Napier (Y-12), James A. Dewar

## Introduction

Technically, "mission-itis" thinking focuses on a particular engine design for a specific mission, with two bad results. First, it produces tunnel vision and hinders seeing a fuller sweep of engine development, as its centers on one design, say with 825-seconds of specific impulse, making it difficult to see that 925- or 1000-seconds can follow quickly. This may not seem important, but 925 means about a 50 percent gain in payload over 825, while 1000 about a 100 percent gain. This turns space flight economics topsy-turvy. Second, it fosters a negative mind-set, to dwell on the problems and believe they will be long, difficult, and costly to solve. This was true in Rover/NERVA's early days where some held a nuclear rocket would need 150 full-scale tests to be man-rated, making the program extraordinarily expensive, but it was only chemical rocketeers trying to impose their experience on nuclear. It was dead wrong - round-peg in square-hole thinking. If Rover/NERVA taught anything, it is that the program always moved much, much faster than anticipated – even stunning its strongest supporters – and a newly reconstituted one can move as fast, if not faster, as it will use new tools, such as CAD/CAM, that the old program never had.

Linked to this is market pull.

Today, it takes about eight to ten years to develop a first-generation chemical rocket engine, and that may cause some to say the prediction of four generations in a decade is hopelessly unrealistic. Not so. That comment derives from the current experience where tax money funds most engine development in industry or government, and that is tied to the annual budget cycle and to very static government or private sector demand. Everything here is slow. Breaking the taboo, however, changes the situation, as it turns the economics of reaching LEO upside down. It creates dynamic market pull that only gets stronger with each better engine. Here NucRocCorp will be set up with competing engine-development divisions staffed with for-profit managers who now operate mostly or wholly with private money. They may start with a first-generation system at $1000 per pound to LEO, but quickly realize the great savings to follow. So now the market will establish demand, not government, and the slow appropriations process, and it will become strong and dynamic if studies show $100 per pound is in the ballpark for a fourth-generation system. With that prospect and competing engine divisions, a new-generation-engine every two or three years is more realistic.

To reinforce this point, consider Pewee; much more advanced than its KIWI and Phoebus predecessors, only nineteen months separated its approval to build from its test in November

1968. And two years later, the even more advanced Pewee 2 sat at Test Cell C waiting for the green light to test when, hit by the buzz saws of environmental impact statements and Nixon Administration funding cutbacks, it was railroaded back to the gigantic EMAD building and disassembled. To further this speed factor, NucRocCorp should establish specifications for each generation engine, and then offer a large cash prize to the division whose engine first meets them. Money changes attitudes, and the prospects of winning the prize and making a lot of money by being first on the market with the best engine creates its own dynamic, and managers who deliver new generation engines in two-three year timeframes will get cash and kudos while those who talk of eight-ten year ones will get coffee and donuts. These points are important as we outline a likely development path of first-through-fourth generation engines.

## II. The Rover/NERVA Experience

Before beginning, it is useful to review key specifications for the NERVA I engine of 1971, which the Nixon administration cancelled: 1500 MW or 75,000-pounds of thrust at a specific impulse of 825-seconds, 60-minutes of continuous full-power operation with 10-hours total and 60-recycles (stops and starts). An upgrade called for 2-hours of full power with up to 12-recycles at a specific impulse of around 850-seconds; this meant increasing the operating temperature from 2100° C to under 2400° C. This was only an upgrade, not a new generation system, as it featured pushing

beaded fuels to their maximum of 2450° C (they melt beyond this), and it would not require engine redesign.

Yet in 1971, the outline of a second-generation NERVA I was visible, with 925-seconds of specific impulse at 2500 MW or 125,000-pounds of thrust. Technically, this meant using the new composite fuels that had just begun development and promised operating temperatures to 2700° C or so. They were tested first in the Nuclear Furnace (a fuel test reactor that appeared in the program's final days), with encouraging results, and Westinghouse tested them with startling results in its electric furnace: 2475° C for over 10-hours and 64-recycles. Thus, a second-generation system at 925-seconds is quite credible. Increasing the power density could be achieved by pumping more LH2 into the core or by making the fuel elements more powerful. The 35-inch NERVA core would have 1870 fuel elements, minus 325 un-fueled ones that contained the tie rod/tube, each with a power of .8 MW. Increasing that to 1 MW per element or 1.1 MW or 1.2 MW thus increases the engine's power. Higher power densities should be possible.

## III. The New Systems

We reviewed that history because some of it is still relevant to the small-large engine progression proposed here, but much of it is not because the earlier experience was dominated by "mission-itis" thinking. Then the overall design always had a nuclear rocket engine starting in LEO and heading deep into the solar system and returning to LEO, with manned Mars the most prominent of those missions. So long run times, with multiple recycles for midcourse corrections and maneuvering, were engineered into them. Here, the flight profile starts at 100,000-feet where the nuclear engine fires and goes several hundred miles up to LEO. Then it returns for reuse. This is the Re-core engine. Slightly later, another small engine would appear and start in LEO and go beyond before returning to LEO. This is the Re-use engine, and the long fuel life and multiple recycles of the earlier era have clear applicability here. After the fourth-generation of this small engine appears, the go-ahead to build a large heavy lifter would be given, and it must be man-rated. A large engine for going beyond LEO would not be developed until later. A final note on the Re-core: it was never studied during the Rover/NERVA era, yet it appears to simplify development. Just remove the old core, recertify the non-nuclear components, insert a new one and fly again.

### A. The First-Generation Systems

We expect the first generation system to be based on Pewee, a 500 MW reactor with a 20x52-inch core tested in 1968, using beaded fuels, which gave a calculated specific impulse around 845-seconds. A flight version would have 20,000-pounds of thrust. Here those who want to make the first-generation better, to incorporate more advances, to do this and that, must be fiercely opposed. What is needed more than anything is flying four years after development starts to give experience: Experience to give confidence to the entire complex that launches and recovers these engines and re-builds

and re-certifies them; experience to demonstrate it can be conducted with the reliability of the airline industry and to show the public the "free launches" will start soon; experience to attract more investors in NucRocCorp; and experience so Congress can herald the new system internationally to attract more countries and peoples to participate. In short, experience to develop and mature the nuclear rocket's infrastructure.

Having said that, there appear to be three major challenges. First, as the engine would startup around 100,000-feet, the nozzle might be somewhat different from one designed for operation in LEO, but this should not be difficult, as a vast amount of data and experience exists on nuclear rocket nozzles. The next two challenges are the engine control system and fuel elements, which are linked in a complex way. The reactor provides the power to drive the turbine that runs the turbopump to push LH2 into the reactor, and different ways exist to do this, each of which feature different ways of siphoning off some hot hydrogen gas from the reactor to power the engine. This bleed-off gas should be as hot as possible to have more energy to power the turbine, but if too much is used, it might subtract from the engine's overall thrust. Then fuel element tie rods/tubes influence this, and their design has many subtleties that allow the hydrogen to be hotter as it returns to the top, cool-end of the reactor. (However, it might be possible to eliminate this totally. See below).

Similarly, the flight profile would influence the control system. With NERVA I, it was to bring the engine to full power in less than 60-seconds, so the rocket would gradually accelerate out of LEO. But going to LEO, that must change; it might take too long, and cause the rocket to lose momentum after the solid rocket boosters stop, and start falling back to Earth. So the objective might be 30-seconds to full power, but this might cause a fuel element buckling problem – the rapid temperature increase could cause the outside of the fuel element to expand at a faster rate than the inside. Yet it is unknown whether this would be a problem in a core to be used only once. Starting the engine after it separates from the cargo plane at 50,000-feet, maybe at 60,000-feet to allow the cargo plane crew to fly safely away, might solve some of this, then having the solid boosters fly a little longer, say to 125,000-feet, might also solve some of this. That probably should take 30-seconds during which time the Re-core engine would come to full power.

We expect the teams to work out solutions to these unknowns in a straightforward manner. As they do so, we expect to see subtle and not-so-subtle differences emerge between Re-core and Re-use engines. For example, a Re-core would vastly mitigate the buckling problem, as fuel has to last for only 15-minutes max, so it would stabilize at one temperature. Repeated heating and cooling wreak havoc on fuel, cracks that open in heating never really close in cooling and when reheated again, expose more of the fuel to hydrogen corrosion. (Hydrogen "eats" carbon out of the graphite at a rapid rate, causing the core to weaken; thus various coatings were developed to protect the fuel element). We also expect to see different engine control systems developed,

one to reach LEO and another for beyond LEO, perhaps with different turbopumps, the axial flow to reach LEO and the centrifugal for missions beyond. The same could be said for the nozzle, one design for reaching LEO, then it may or may not be discarded. Since it has to last for only 15-minutes max, its economics of manufacture might become a dominant factor. Nozzles destined for re-use to and from LEO would see reliability and ruggedness engineered into them, making them last for as many missions as possible. That would raise their costs and be more difficult technically, as nozzles should be constructed with thin-walled tubes to permit a rapid heat transfer, but thin-walled tubes are inherently not rugged.

## Note on the Cocoon for Engine Design and Operation (James A. Dewar)

Rover/NERVA never had a flight-rated cocoon; the engine was to start in LEO and go outward, so it, from side to side, had a metal pressure vessel, then a beryllium reflector holding the control drums and finally the core, and from front to back, the order the components were placed was quite standard, as the NERVA illustration indicates: LH2 tank; turbopump/turbine; pressure vessel and core; and nozzle. Engine weight was also important, with the desire to keep it as light as possible. Here weight is not so critical a factor and the cocoon's dimensions, its size and shape, would be limited principally by the large diameter of the LH2 tank. In other words, it could be very spacious, and have different shapes – a re-entry vehicle for a warhead, an Apollo-like capsule, a lifting body, and a clamshell - and that room can become a key part of engine design and operation. For instance, the metal pressure vessel might be eliminated totally, leaving the cocoon to hold the core; the beryllium reflector might be thicker, which in turn will affect the core's neutronics, perhaps making criticality easier for the small Re-use and Re-cores; the use of poisons (neutron absorbing materials) throughout the cocoon might be widespread to dampen the radiation, eliminating the need for the shield; the turbopump/turbine might move inside the cocoon, or remain outside it; and the power to run the turbine might change to an auxiliary system in the cocoon. (As noted in NERVA, the engine was to power the turbine/turbopump, but this siphoned off up to 5 percent of the engine's thrusting power). In sum, a cocoon offers engine designers many options not allowed with metal pressure vessel systems. They probably would devise a simple cocoon for a first-generation engine, but later-generation ones would likely have more sophisticated ones with features to improve performance. These have never been studied, so it would take time to sift through the possibilities, but I hold they would play a critical role in engine design henceforth. I'll talk more about the cocoon in the context of flight safety in Appendix C.

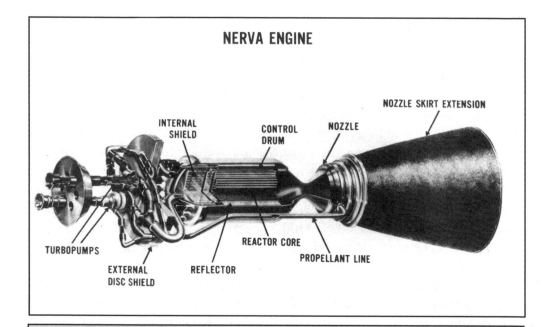

**NERVA ENGINE**

**NERVA**

PHOTO CREDIT: NASA

This exploded view of NERVA illustrates the arrangement of components. A cocoon enveloping the reactor will allow designers to question this order and some parts might be moved to different areas in the cocoon or outside it, such as the turbine/turbopump, and some might be eliminated, such as the metal pressure vessel. The cocoon would become the pressure vessel.

## B. The Second-Generation Systems

We assume NucRocCorp would realize fuel element development was the most critical item of the solid core, and take great pains to ensure it was properly organized and supported. This has been mentioned before, but no harm comes from stressing it again: It is truly that important. At a minimum, there should be at least three different groups developing fuels, in competition yet in full cooperation with the others, each with its own electric furnace, and *one or more* small reactors to test fuels as desired. This follows a key lesson from Rover/NERVA: Electric furnace testing became less useful as temperatures and recycles increase; in other words, a radiation environment was mandatory. Such testing need not be conducted at a test cell at Jackass Flats but follow the Russian experience where it is tested indoors, without worrying about weather conditions (or U.S. spy satellites). A facility such as the giant EMAD building at Jackass Flats could be reconfigured to allow it, and since these test reactors would have scrubbers on them, as the original Nuclear Furnace did, any worry of environmental hazards must be dismissed. Moreover, EMAD's hot cells would allow for on-site fuel post-mortems quickly. No doubt other DOE sites could also be configured to test fuels; it is not facility intensive. While beaded fuels would be used in the first-generation system, composites would be used thereafter. If a first-generation system appears four years after the program restarts, a second-generation should follow two or three years later, or about six or seven years after the program began. This gives ample time to spend on composite fuels.

In 1971, composites were thought to have the potential to operate around 2700° C and we assume our second-generation system will operate around 2600° C, an increase by 400° C to give a specific impulse around 925-seconds. This will cause an increase in the stresses and strains the engine would undergo. Everything would be much higher: the radiation levels, the operating pressures, the temperature, the vibrations, and so forth. This would require an engine upgrade, but these would be engineering problems, not barriers. The nozzle, the pressure vessel, the engine control system, the turbopump, the actuators (motors that rotate the control drums) and the few other parts that make up an engine could be designed and developed to handle this new operating regime. In fact, we categorically state they could be developed for any nuclear rocket-operating regime, including 3000° C+, and will not discuss them further.

That leaves the fuel and fuel element, and here we see the first real division between the Re-core and Re-use engine, with the former having the propellant hole enlarged. Now Rover/NERVA fuel mostly had a 0.10-inch propellant hole. Here for a second-generation Re-core that might be enlarged to 0.11-inch, a hundredth of an inch, a seemingly minuscule change. (Remember each fuel element was three quarters of an inch from flat-to-flat and had 19-propellent holes in it). This larger size, seen across the core's diameter, increases the amount of room for LH2 to pass through; in other words, it allows more LH2 to be pumped into the core, so power density increases. This also creates a thinner web between the holes, and that should allow the fuel to increase in temperature more rapidly without buckling better than a thicker web. Maybe we're around 600 MW now. But it's not quite that easy. Thinner webs would mean less uranium is in the core, and this would compound the criticality problem of the 20-inch core. The original Pewee had 534 fuel elements, but only 402 loaded with uranium; the remaining 132 were tie tubes made out of a moderating material to promote criticality. So making the propellant hole a hundredth of an inch larger could make criticality more difficult, but ways exist to work around it, such as increasing the density of the fuel and increasing the thickness of the reflector. All involve trade-offs, but we expect a conservative attitude to prevail, and an over-engineered system to be the result.

The Re-use engine would probably have the propellant holes at 0.10-inches, as recycling would be emphasized, and here we expect the fuel to last 15-hours at full power, 90-recycles at 2600° C. This would a reasonable extrapolation, given five, six or seven years of development, particularly since Westinghouse had 10-hours at 2475°C and 64-recycles in electric furnace tests in 1972.

## C. The Third-Generation Systems

These would appear eight, nine, or ten years after the program was restarted. We would expect both Re-core and Re-use engines to undergo a major refinement as by now the development teams would have ample experience, and that intuitive knowledge of how to make even better engines. We expect they would operate in the 2800-3000° C range, several hundred degrees higher than the second-generation system, and seemingly beyond the limit of the composite fuels, as conceived in 1971. However, back then Y-12 was investigating adding uranium to commercially available ion-exchange resins to make the fuel particles, which appeared stable in tests at 3000° C for an hour. This was startling because the published literature indicated it should melt well before this, and apparently the increase was due to sulfur impurities in the resin. So back then that was among the things Y-12 was considering to improve the fuel, thus increasing the range of composites to around 3000° C. That promises specific impulses around 1000-seconds. Their upper limit now appears higher, but how much is quite uncertain, as graphite, a main ingredient in composite fuels, sublimates (turns from a solid to a gas) at 3700° C.

So third-generation Re-core and Re-use engines should have around 950+ seconds of specific impulse, but the former might have the propellant hole increased a bit more than 0.11-inches though the web would be getting pretty thin now. The Re-use might see an increase to 0.11-inches. This would mean different power densities and therefore different power levels for each engine: For the Re-core the progression might be 500 MW, 600 MW and now 700 MW or higher; for the Re-use, the progression might be 500 MW, 575 MW, 650 MW. We also expect continued emphasis on increasing fuel density, as criticality would remain a problem with these small-core engines, though adding more reflector material to the cocoon could change that. And in increasing several hundred degrees in temperature, unexpected problems are likely to develop that would take time and creativity to solve. This would be particularly true for long-life, multiple-recycle fuel, so 20-hours and 120-recycles might be reasonable to assume in almost a decade of development. But it still would be hard.

If the second-generation engines were over-engineered, refinements should occur with the third, such as making them lighter or giving them greater operating flexibility. A key change might be the reduction in the size and shape or the total elimination of the shield between the top of the engine and the turbine/turbopump/LH2 tank. In Rover/NERVA considerable attention was given to this juncture and the shield – its size, shape and thickness – because everyone feared radiation would turn LH2 into a gassy-sluggy liquid *before* it entered the turbop-

ump and cause the pump to stall out and not push any hydrogen into the engine. Once LH2 left the turbopump, it was pressurized and heated and turns into something like steam (water heated above 100° C under pressure). Henceforth, it was not a problem. Back then, however, Rocketdyne demonstrated the turbopump could still pump LH2 even if a third of it was gaseous. This meant the shield could be reduced in size or eliminated totally, thus saving a lot of dead weight. The first-and-second generation-systems would likely have shields, but a strong component effort likely would have been underway since the program began, and would likely see fruition here with components able to handle the higher radiation levels. However, the cocoon could make the shield unnecessary, as it could be filled with poison materials that dampen the radiation.

Another likely change would be the introduction of the Re-core Long Fire engines. Developing fuel to run for 15-minutes max is straightforward, but undoubtedly the divisions would look at the NRX-A6 that ran for more than hour at full power in 1967 and ask if a market existed for a Long Fire engine, ones that startup at 100,000-feet but fire continually for an hour or more to propel a payload on a one-way mission into the solar system. These might not be economical at 825- or 925-seconds, but closing on 1000-seconds of specific impulse could change the situation, even if the economics include the cost of the highly enriched uranium. (These might be government missions, so the cost would be waived). The large LH2 tank size and the cargo plane's capacity would prevent Long Fires with first- and second-generations systems, but with a decade's experience the situation will change. Perhaps then large cargo planes might be designed to carry them, and perhaps all launch divisions might now be using slush hydrogen (below), meaning the tank size for a given volume of LH2 might be up to 20 percent less. Moreover, the solar system transportation plan should be fairly developed by now, and we would not be surprised to see some divisions starting to specialize in engine development to fill in the niches, such as dual-purpose engines – ones that provide propulsion power and electricity to the payload or nuclear ion engines.

### D. The Fourth-Generation Systems

These would appear about a decade after the program started, but now our discussion becomes more difficult. We expect continued progress on composite fuel, but can't say how with confidence. We can only say that research and development produce many surprises, and many ways to solve problems that were not apparent beforehand. Based on current knowledge, it appears composites reach their technological limit somewhat higher than 3000° C, meaning Re-core and Re-use engines would have specific impulses around 1000-seconds, perhaps slightly higher. Even that, though, might change to a higher number. We also expect improvements in other areas such as making fuels denser, making

the propellant holes a tad larger, or having more of them in the fuel element (once a 52-hole fuel element was considered), increasing its operational life and number of recycles. Here, progress may be slow and difficult; nonetheless, it is possible. So we assume the Re-core will be 800 MW in power with the Re-use perhaps slightly less, with 1000+ seconds of specific impulse. This means about 20,000-pound payloads. However, we advance the idea of 30-hour fuel lifetimes with 180-recycles with some trepidation. Now it appears quite unattainable, but with at least three different groups working on composites for a decade, it could be the norm, with better fuels to follow.

### IV. Transition to the Larger Engine

Once approval to build large engines is given, perhaps ten years after the program began, the first engine should appear two or three years after that, followed shortly by even more powerful model with 3000 MW of power and 1000-seconds of specific impulse, able to propel 150,000-pounds to LEO. In other words, we expect a jump from a first-generation to fourth-generation system quickly, without the earlier step-by-step approach. Some may hold this as too optimistic, but bear in mind the competing divisions would develop them on a schedule not determined by the government budget cycle but by market pull and money coming from private sources. This would be a Re-core heavy lifter, as missions for a large Re-use engine will await more zones opening for development, and most likely it could increase even in power, as the 35-inch core does not have the criticality problems of the 20-inch core. In other words, it could be even larger than 3000 MW yet still in the 35-inch core, meaning even more payload to LEO, so the 150,000-pound estimate might be too low. Also, core sizes other than the 35-inch could evolve to satisfy different heavy lifter markets: 40-inch, 45-inch, 50-inch or even 55-inches, but they would require their own launch vehicles. The non-nuclear components – the nozzle, turbopump, actuators, and engine control system – would be straightforward to develop, given the decade-plus experience with the small engines. Finally, with the cores operating at 3000° C or higher, it opens up the possibility of using other propellants than hydrogen for special applications; those would include ammonia, methane and water.

## V. The Wild Cards: What We Have Excluded from the 20,000/150,000-pound Payloads

There are four wild cards – ideas that seem workable and could improve the capability of the Re-core/Re-use engines markedly, thus lowering the costs of moving payloads into and beyond LEO. These could become feasible when fourth-generation engines appear, or their appearance might not occur until later-generation systems.

### A. The First Wild Card

We have not included the carbides, the metals that melt around 4000° C and promise specific impulses well over 1000, even perhaps to 1200-seconds. These strong, but brittle fuels just started development in the Nuclear Furnace test and had a promising future though to use them effectively requires a new core concept. The B-4 won't do. However, Rover/NERVA had many afterburner concepts, some of which featured replacing the last foot of a 52-inch long fuel element with a carbide element, thus boosting the specific impulse well over 1000-seconds. That, of course, would affect payload numbers and economics. Yet it may turn out their brittle nature makes them ideal for a new design Re-core engine, since it would fire only once, and if so, super-lifting small and large Re-cores might appear with specific impulses perhaps close to 1200-seconds. This appears startling and unrealistic, but a decade ago a Russian newspaper reported its nuclear rocket program tested carbide fuels at 3200° C for an hour. That is unverified, but we know the Russians mysteriously switched their nuclear rocket program from graphite to carbide fuels in 1968, and henceforth tested many varieties for hours before running into funding difficulties in the 1990s. So given their experience, it's not out of the realm of credibility. Running around 1200-seconds could have stunning consequences for the economics of reaching LEO; going from 825 to 1200 promises almost a quadrupling of the payload. This is partially because of hydrogen disassociation, its molecule splitting into its two atoms. Disassociation starts around 2200° C and is completed around 5000° C. Think of a bell curve. As 3200° C is almost mid-way on this curve, disassociation could boost performance noticeably. Atoms would now be ejected, not just molecules, so weight has been lowered again. Moreover, radiation weakens the bond of hydrogen's two atoms to each other, so a new line of research might be to see if the hydrogen flowing from the turbopump into the nozzle and engine could be weakened even further by a radiation source so even more hydrogen atoms would be ejected out the nozzle.

## B. The Second Wild Card

We have not included the weight of the stage – the LH2 tank is large and heavy as LH2 is not dense. A cubic foot of it weighs about 4.4-pounds - a cubic foot of water about 62-pounds. However, as Rover/NERVA ended, research into slush hydrogen (SH2) was beginning; SH2 is LH2 cooled past -252° C to where it condenses into a slushy form – think of a snow cone. As it does, it shrinks anywhere from 15 to 20 percent in volume. This means a corresponding reduction in a tank's weight for a given amount of hydrogen, or more hydrogen for a mission. Hence, the tank's weight could decrease by up to a fifth, saving thousands of pounds that now could become payload; with the small Recore with its tank estimated at 20,000-pounds, that could be up to a 4000-pound savings. The principal unknown is whether the turbopump could handle SH2. We believe it could, but expect a development effort to prove it. Thus, when SH2 might make its appearance is unknown. Certainly, it would not be for a first-generation engine, and it has been excluded from the payload numbers used for the fourth-generation small engines. (Actually, triple point hydrogen appears better than slush, as it shrinks to about 90 percent of slush's volume and wouldn't be as hard to handle; pure slush appears to require a mixer in the tank, adding needless complexity and reducing reliability).[1]

## C. The Third Wild Card

We have not included tie tube development. In 1955, Los Alamos picked graphite as the core material, as it had good neutronic properties and grew in strength until it sublimated at 3700° C. For 1955-era graphite, however, that strength was only in compression or when pushed together. When in tension, when pulled, it was weak and brittle and broke readily. Thus evolved the B-4 core, with its tie rods initially and tie tubes subsequently, to hold the core in compression. Rover/NERVA led a fundamental assault on understanding graphite and developing better varieties, also called carbon composites or graphite fibers. By the early 1970s, Y-12 had varieties with good properties in tension and had just begun to investigate using them to replace the metallic tie tubes when the program was cancelled. However, Y-12 continued to work on carbon composites for other purposes for the next several decades, so we expect this effort to be restarted in any new program, as it promises a major increase in power density, perhaps to the all-loaded core, and it could ease the neutronic properties of the 20-inch core and allow a much higher operating temperature. This means the 800 MW and 3000 MW power levels we predicted for fourth-generation small and large engines might be on the low end. So again, the payload numbers might be too conservative.

### D. The Fourth Wild Card

All research and development programs invariably lead to unanticipated and unforeseen ideas that improve the product, but when and how they occur is impossible to predict. This is particularly true for the cocoon as it relates to engine/core design and operation. It would be naïve to assume such would not happen here, but *when* is the question. If before our norm of the fourth-generation engine, the payload improves; if after the norm, the payload also improves. When Rover started in 1955, some predicted specific impulses of 1600-seconds from the solid core; that seems utterly unrealistic, a pipe dream, and even 1200-seconds seems to be pushing the outer edge of the envelope. Yet with different fuel element groups in friendly competition and engine development divisions in fierce, market-driven ones, it would be unwise to state 1200-seconds was the upper limit.

### VI. Summary

The 20,000- and 150,000-pound payload postulated in the text for fourth-generation engines suffice for the policy argument of this book. Actually, if Rover/NERVA's experience was duplicated here, they would be probably the lower limit of what would appear a decade after the program restarted, as the earlier one always moved faster than even its most ardent supporters thought possible. So we urge the reader not to get too hung up on them, as they are offered only to illustrate a much fuller sweep of engine improvement to offset static "mission-itis" thinking, and provoke debate and dialogue over a new way of thinking about nuclear rockets. Hopefully, studies will appear shortly with much better numbers, as the new thinking challenges those in the space community.

For those seeking more information, we recommend the following:

1. F. Durham, Nuclear Engine Definition Study Preliminary Report, LA-5044-MS, Los Alamos National Laboratory, September 1972.

2. Stan Gunn, Design of Second-Generation Nuclear Thermal Nuclear Rocket Engines, AIAA 90-1954, July 1990.

3. Stan Gunn, James Edstrom, Rolf Honda, Power Generation Capabilities of Tie Tube Assemblies, Eleventh Symposium on Space Nuclear Power and Propulsion, Albuquerque, NM, January 1994.

4. Stan Gunn, Design of Superheat Nuclear Thermal Rocket Engine, AIAA-94-2895, June 1994.

5. J. Napier, F. Homan, C. Caldwell, Particle Fuels Technology for Nuclear Thermal Propulsion, AIAA 91-3457, September 1991.

6. Candidate Nuclear Fuels for Nuclear Thermal Rocket Propulsion, na, nd, (author's files).

---

[1] Per Norman Gerstein.

# Appendix C
## A Walk-Through the Re-core's Flight Profile to and from LEO

### Introduction

Here I walk-through the Re-core's flight profile to discuss its hazards, and outline how they could be overcome or lessened. This never-considered topic is complex and involves many subjects, from Re-core design and operation to range safety and launch corridors to accident recovery. To simplify my discussion, I group them into ten-steps to show this flight profile would be doable, and would have a level of risk equal to that of the airline industry. That seems an apples and oranges analogy, at least initially, as airlines have experienced one fatality per two billion passenger air miles flown since 1997 while the Re-core would be unmanned until heavy lifter engines appear. Yet the common thread is very little risk; flying is the safest way to travel and, I hold, heavy lifters would make flying into Space for mankind just as safe when they appear.

### I. The First Step: The Cocoon That Envelops the Re-core

The previous appendix sketched how the cocoon could be integrated into engine design; here I review it in the light of launch and recovery operations, and accidents. I repeat Rover/NERVA never had cocoons. Only the pressure vessel was to stand between the reactor and the environment, as depicted in the illustration, so the gamma rays and neutrons radiated out in varying degrees of intensity before dissipating in Space. Looking closer at the illustration from the 1960s, however, reveals the NERVA engine was to be 3-feet wide while the LH2 tank 33-feet in diameter. That wide space is similar for the Re-core; the engine would be 2-feet wide and the LH2 tank 13-feet in diameter (allowing the stage to fit inside a C-5A's cargo bay). This is critical, the 13-feet give great freedom in designing a cocoon, so it could be buoyant, even with a 6000-pound engine, as well as survive re-entry without the release of radioactivity, and it would have ample room for poisons, such as boron, to dampen the neutrons and gamma rays (those wavy lines in the illustration), black box recorders, telemetry instruments, accident prevention or mitigation equipment, parachutes and extra flotation gear. Its actual shape could be like the Mk-6 re-entry vehicle for the W-53 warhead, an Apollo/Orion-like capsule, a lifting body, or a clamshell cocoon. I expect a first-generation cocoon would be fairly simple, but be upgraded as experience was gained. Whatever its shape, it will be a big first step in eliminating or mitigating hazards, thus reducing risks.

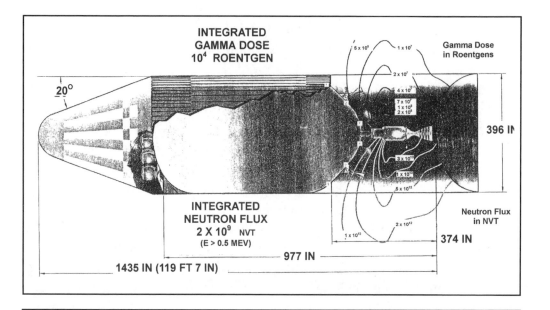

INTEGRATED
GAMMA DOSE
$10^4$ ROENTGEN

Gamma Dose
in Roentgens

$5 \times 10^6$   $1 \times 10^7$

$2 \times 10^7$

$4 \times 10^7$

$7 \times 10^7$
$1 \times 10^8$
$2 \times 10^8$

$3 \times 10^{12}$

$1 \times 10^{12}$

$5 \times 10^{12}$

20°

396 IN

INTEGRATED
NEUTRON FLUX
$2 \times 10^9$ NVT
(E > 0.5 MEV)

Neutron Flux
in NVT

$1 \times 10^{12}$   $2 \times 10^{12}$

374 IN

977 IN

1435 IN (119 FT 7 IN)

### RIFT DOSE CHART

CREDIT: NASA

This shows the neutron and gamma ray dose projected from a NERVA propelling a huge Saturn V stage 33-feet in diameter. It had no cocoon. A cocoon enclosing the Re-core would contain poisons such as boron or radiation shielding to dampen the radiation throughout the flight profile, from launch to recovery.

### MK-6 RE-ENTRY VEHICLE

CREDIT: NATIONAL ATOMIC MUSEUM

Sitting atop the Titan ICBM, the Mk-6 re-entry vehicle was 10-feet tall, 5-feet wide and weighed 2100-pounds. It contained a W-53 9 MT warhead that was 8.5-feet long and 3-feet wide and weighed 6500-pounds. The Mk-6 might bear a close resemblance to a cocoon developed initially for the Re-core.

### LIFTING BODIES

PHOTO CREDIT: NASA

This is the M2-F3 lifting body being dropped from a B-52 in 1971, demonstrating the ability of pilots to maneuver a wingless vehicle and land safely from space like an aircraft at a pre-determined point. A Re-core might be enclosed or encased in a pilot-less lifting body for its return from LEO and land in an isolated area such as island in the Pacific. Landing pilot-less vehicles is common now.

### ORION RE-ENTRY CAPSULE

CREDIT: NASA

The Orion re-capsule is being developed for the Ares I and is similar to the Apollo capsule. This shape could be designed to house a Re-core engine for re-entry. The Re-core would be located in the center, with the top of the engine facing the heat-resistant mushroom  and the nozzle protruding out the more pointed portion.

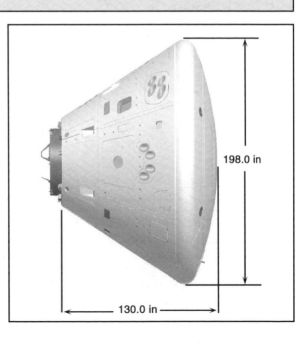

198.0 in

130.0 in

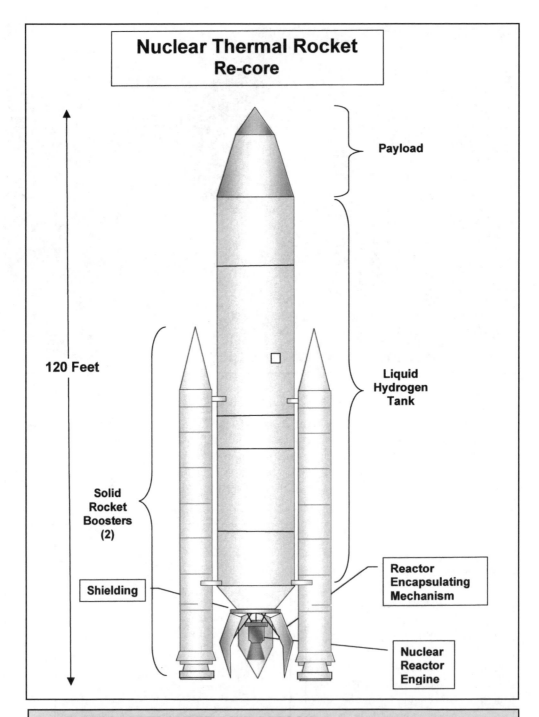

**Nuclear Thermal Rocket
Re-core**

Payload

120 Feet

Liquid
Hydrogen
Tank

Solid
Rocket
Boosters
(2)

Reactor
Encapsulating
Mechanism

Shielding

Nuclear
Reactor
Engine

**RE-CORE STAGE AND RV-LIKE COCOON**

In Chapter 5 a Re-core stage with re-entry vehicle-like cocoon was illustrated; here a clamshell-like cocoon holds the engine. Orion/Apollo-like and lifting body cocoons are also possible. Each, however, presents its own unique design challenges where the cocoon joins the LH2 tank, in particular for the line feeding LH2 from the tank to the engine; for the gimbal that keeps the stage flying straight; and for the structure to link the cocoon to the LH2 tank/stage.

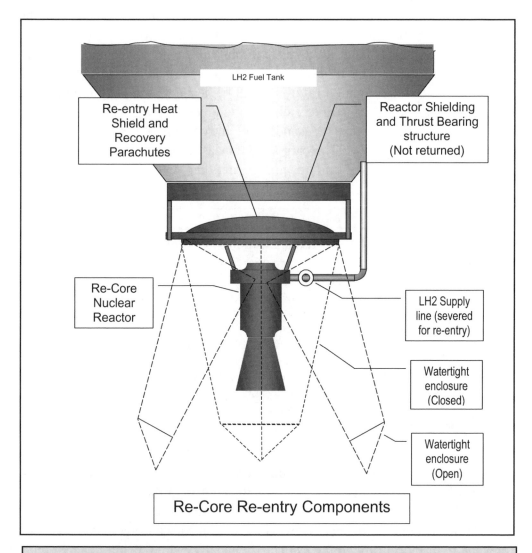

Re-Core Re-entry Components

**CLAMSHELL COCOON**

This clamshell cocoon concept traces its lineage to the clamshell shields developed for the XE-1 test in 1969. The two never sealed properly, but then they were much larger and needed to be railroaded into place. These much smaller clamshells might make it easier for this concept to work.

## II. The Second Step: The Launch Corridor

Our walk-through continues with the launch complex and range safety. Chemical rockets normally require permanent launch installations such as Cape Kennedy, so they have fixed launch corridors from which the flight into Space occurs and in which a range safety officer decides whether it is go or no-go, and pushes the self-destruct button to prevent an errant rocket from landing in populated areas. On more than one occasion, a rocket has been destroyed even though it would have performed its mission because it gave indications it might stray outside these protected boundaries. The Re-core's flight profile changes that; it needs no fixed installations; its flight corridor is moveable, and many are possible, such as in the vast Pacific Ocean, a thousand miles from populated areas. The cargo plane just flies from an airport to an isolated area there and launches. So the second step would require totally rethinking range safety and the flight corridor. I suspect range safety experts will be hesitant at the beginning, but a growing familiarity with the Re-core and its cocoon will make them wax enthusiastic, as they see safety margins greater than with chemical rockets.

---

### Non-fixed Launch Corridors

The non-fixed launch corridor, like that presented here, is not new. Rather, it's a staple of the U.S. nuclear deterrent; nuclear submarines that can launch their missiles from anywhere at sea, even from under the North Pole. However, for commerce, its roots go back to Project Argus and the ship Norton Sound, which fired three rockets to 300-miles up in 1958 in the South Atlantic Ocean, 1000-miles east of South Africa. Today, Sea Launch, an international partnership that includes companies in four countries (United States, Russia, Norway and Ukraine), is employing it. It is a proven technique for launching large rockets weighing in excess of a million pounds, launching them from isolated areas near the equator. The possibility also exists for airborne launches as with Orbital Sciences Corporation's Pegasus.

---

### III. The Third Step: Preventing Accidental Criticality

The next step is criticality: How to prevent a reactor from going critical (from sustaining a chain reaction) when we don't want it to in launch, accident modes, or re-entry. I'll discuss this throughout this appendix, but it begins with reactor design. Nuclear engines are designed to be sub-critical (as are all reactors), meaning one could stand next to them without getting a harmful dose of radioactivity, as the Phoebus 2A photo shows. They go critical only when LH2 enters the core and/or the control drums that regulate neutrons in the core are turned open. Then one cannot stand next to them. This sub-critical design means a Re-core can be handled manually while being loaded on the cargo plane and flying to an isolated area. At a minimum, the control drums would be locked into place to prevent criticality and unlocked only when dropped to begin the flight to LEO. This was one technique planned for the RIFT (Reactor-in-Flight-Test) of the early 1960s; this area will receive further study, and no doubt other methods will be found.

**PHOEBUS 2A**

PHOTO CREDIT: LOS ALAMOS NATIONAL LABORATORY

CAPTION: The Phoebus 2A had a design power level of 5000 MW in a core 55-inches wide by 52-inches long, making it the most powerful reactor ever built in the world, even today. This photo shows technicians preparing the cold reactor for hot testing in 1968, and illustrates how a Re-core can be handled manually to mate it to a cargo plane/C-5A for its flight to an isolated ocean area for launch without radiation hazards to personnel.

## IV. The Fourth Step: Reliability

The fourth step is reliability. We know its importance implicitly: Our automobiles, telephones, airplanes, and railroads all operate with near perfect 99.99 percent reliability, and we expect it. When accidents happen, which they certainly do, the cause tends to be more human error than technological failure. We tolerate them, but want safety improvements. With this flight profile, the reliability of the cargo plane such as the C-5A, the solid rocket boosters, and the Re-core will be a key factor in any risk assessment. Since introduced in 1969, the C-5A has had a remarkable safety record given the millions of miles it has flown, with about ten major accidents, most of which occurred during landing or on the ground during maintenance operations. If it cannot be used, a specially designed cargo plane would be built and likely have the same reliability, as this technology is quite mature now.

**C-5A**

CREDIT: AIR FORCE

The C-5A entered service in 1969 and was last produced in 1989. It has a maximum pay-load of 291,000-pounds and has a cargo bay 143-feet long, 13.5-feet high and 19-feet wide, which could hold the Re-core stage. That is not the problem; whether it could launch the stage at 50,000-feet out its rear, via a drogue parachute or similar device, is. If impossible, the next option is modifying the C-5A to do so; this might feature dropping the stage like a bomb. If impossible, the next option is design a cargo plane specifically to launch the Re-core stage. After studies, that might end up the best option.

Next, solid rocket boosters have reliability greater than 99 percent. The Re-core's would be at least 99.7 percent - the NERVA I was to have 99.7 percent reliability - and it was to be more complex than a Re-core. Stated differently, this means 997 times out of 1000 the nuclear engine would operate as designed. Statistically then, there is an overwhelming probability the C-5A/cargo plane, the solid rocket boosters, and the Re-core would operate as required in the flight profile. This reli-ability factor does not necessarily mean an accident causing the loss of the stage, only that the engine did not operate as designed.

### V. The Fifth Step: What Happens If? Three Scenarios

With four steps taken to prevent or mitigate accidents, what happens if one still occurred during launch? What happens if the 3 out of 1000 were not minor abnormalities but accidents causing the loss of the stage? That's the fifth step, and it's a fair question; the answer is "It depends." I'll answer by saying the stage consists of the engine/cocoon, the LH2 tank and payload, and solid rocket boosters, all of which can be separated from each other with minimal risks to the flight crew or the environment, so that leaves the cocoon to worry about. Thus, the severity of an accident depends, perhaps most importantly, on the height at which it occurs. There appear to be three major scenarios: First would be a failure to drop properly from the C-5A/cargo plane at 50,000-feet. Second would be a solid rocket booster failure from 50,000 to 100,000-feet. Third would be Re-core failure from 100,000-feet to 200-miles up. Common to all three is the cocoon plummeting back with a cold core (first scenario), with a slightly hot one (second scenario), or hot one (third scenario), and landing in the ocean or isolated land area.

The first line of defense here is the engine itself; it is almost as rugged as an armor piercing shell, so it *alone* would tend to return in one piece. That makes recovery easier. The Soviet space power reactor that crashed in Canada in 1978 came down in big chunks, making cleanup easier. A Re-core with a metal pressure vessel would be even more rugged than it. The second line of defense would be the cocoon. It would be designed to handle the rigors of re-entry and withstand any accidents that might occur during the 50,000-foot to 200-mile

high flight and return to Earth with the core intact. Though the cocoon might be damaged, it would contain the core, and the accident prevention or mitigation equipment would limit or prevent the release of any radioactivity..

That's a bold statement but to back it up, I look to Rover/NERVA for insight. In early flight safety tests, a mockup reactor was dropped from 75-feet to a concrete pad, and another slammed via a rocket sled into a concrete wall at several hundred miles per hour (doing that with a chemical rocket engine yields scrap metal). In the first, it bounced 15-feet high, denting the pressure vessel; some fuel elements cracked, but calculations indicated the core would not have gone critical. In the second, the core compressed 5 percent; calculations indicated there would have been an explosion equal to about 4-pounds of high explosive, with the core going critical for a split second. This rocket sled test might be a useful point from which to begin studies, as it might approximate the force of an engine/pressure vessel hitting the water in an accident. Now, however, we surround the pressure vessel with a cocoon, adding not only another layer of defense but also gear to prevent or mitigate any accident, the most important of which would be to scram (an emergency shutdown) the reactor. This might occur if the flow of LH2 to the core was interrupted and start by rapidly turning the control drums to "off," then going to a cocoon-located emergency core cooling system to remove excess heat from the core.

Rover/NERVA planned pulse cooling, sending puffs of hydrogen with the drums off to reduce the heat, done because NERVA was to be reused, so maintaining the core's integrity was important. That's not important here, so designers have wide latitude. A solution might be the cocoon containing helium (widely used in cool-down in Rover/NERVA) or a poison-containing gas to reduce heat and prevent criticality. Another might be a plug that slides into the nozzle. A third might be a fast acting poison cement that floods the core quickly, then sets up and hardens. So here the actual risk would depend on many factors, primary of which is the cocoon's design to contain accident prevention or mitigation gear.

With these perspectives on the cocoon, let us examine the first scenario where the C-5A/cargo plane is taking off and flying to 50,000-feet. Our first concern is the LH2 tank leaking, and hydrogen igniting, a potential fireball, a Hindenburg dirigible disaster. However, one way to ease that worry is sealing the cargo compartment and flooding it under positive pressure with an inert gas, such as nitrogen. No oxidizer, no fire, no explosion. So we continue: Something happens after the stage leaves the C-5A/cargo plane, the solid boosters fail to start, and it starts falling back to Earth immediately. As the Re-core is cold, it has not gone critical; it would be separated from the stage, and the criticality prevention gear activated. Then it would splash down in the ocean and be picked up. There is no radioactivity hazard to recovery crews. The solid boosters would pose little hazard as they splashed down, but the LH2 tank raises the possibility of a fireball since the accident would occur at 50,000-feet where enough oxygen may be present to permit combustion and perhaps an explosion. The payload might be recovered, but that depends on its design. This scenario requires further analysis, but its risks appear to be quite manageable; the public would not be affected, even if a LH2 fireball occurred, because the accident would occur thousands of miles away from populated areas.

Now I'll look at the second scenario where we're between 50,000-and 100,000-feet and somehow the solid boosters malfunction as the Re-core was coming to full power. So it's slightly hot. What happens then? The simple answer is we abort the mission, scram the reactor, activate the accident mitigation gear, and then pick up the cocoon in one piece. Here is a major change with chemical rockets. If one becomes errant, the range safety officer pushes the button and blows it up, the pieces scattering over the ocean floor, the astronauts hopefully returned, the payload perhaps recovered, and a promise to do better next time. Not so here. The LH2 tank can be breached to dump this flammable fluid, but going ever higher than 50,000-feet means ever-less oxygen is present, except that from the solid boosters. Again, no oxygen, no combustion, no explosion. The super-cold LH2 just boils away and disperses to the rarified atmosphere. So a blast and fireball common to chemical rockets becomes increasingly impossible (yet another reason range safety experts will love Re-cores). The payload, unscarred and un-scorched, might survive, but this depends on its design. I expect when heavy lifters appear, the bus carrying 50 people would be designed to survive and here is the thread back to airline safety risk levels. If an airliner disintegrates in flight, all are lost; if a heavy lifter disintegrated, the bus could be designed to return its passengers safely to Earth. Now the cocoon, also unscarred by an explosion and unscorched by fire, would return and must be picked up, but it would be slightly hot, as it would have run for maybe 30-seconds. This short time is sort of like the KIWI-TNT test, in which a full-size reactor was engineered to blown up and it did with a force equal to 300-pounds of black gunpowder. Afterwards, crews picked up the pieces with tongs. So, in principle, this second scenario should be quite manageable and cause little harm.

Now I'll look at the hot scenario where the Re-core malfunctions between 100,000-feet and 200-miles up. Here, the solid rocket boosters have been jettisoned and the Re-core is at full power when something happens to it. This is the cripple or KIWI-B1B scenario, and the first question a range safety officer would probably ask is "Do I still have control?" If no, then the mission is aborted, the reactor scrammed, the accident gear activated, and the recovery teams given a go-ahead. If yes, it is similar to the KIWI-B1B test, which ejected the core out the nozzle, but it was always controllable and kept churning out power. This implies if the Re-core could carry out the flight to LEO. Or if it couldn't achieve a stable orbit, it might fly until it reached a specific area for recovery. This would be particularly important for heavy lifters propelling buses: Unlike errant chemical rockets that give the range safety officer little choice even with astronauts on board, the Re-core heavy lifter gives him options before he pushes the button: Keep the cripple flying till it reaches LEO, till it's over an isolated area, or until the bus has been safely separated from the stage, then push. (This is another reason why I believe range safety experts will come to love the Re-core).

Let's assume the range safety officer decides to fly the cripple to LEO. What are the consequences of an engine spitting out fuel elements at high elevations? In general, if ejected above 12,000 mph, most would burn up during re-entry, and the residue would disperse at high elevations. Based on data from space nuclear power sources returning to Earth, this radiation would be measurable until it dispersed to levels where sensitive instruments couldn't detect it. At a speed of 8000 mph or less, they would drop into the ocean and sink to the bottom. In both, the radiation decay process would begin at once to lower levels. What would be that impact?

Here I look to the Phoebus 1A accident for insight. It was hot after 10-minutes of full power operation, the same length of time the Re-core would be, and it ejected a third of its core. Yet after six weeks it had decayed to the point where it took 400 men armed with tongs and a vacuum cleaner to clear up the mess at a cost of $50,000. This suggests not much of a lasting impact.

While these are simple answers, more realistic ones are nuclear engines will require an emergency response infrastructure, and here experts would analyze all plausible accident scenarios, and the best response to each. All this, ultimately, leads back to the cocoon. A floating, buoyant one is possible even with a 6000-pound engine. If the cocoon survived re-entry as designed, it would splash down somewhere in its flight corridor. So now the floating cocoon must be picked up and here the ocean water would give radiation protection for recovery crews. The principal exposure would occur in lifting the cocoon out of the water and into a shielded cask on deck, an operation taking only a short time. Yet if the cocoon somehow failed and the engine sunk to the ocean floor, plans would also include deep ocean submersibles to recover it. In developing this data, a good place to start is the Glomar Explorer, which in the 1970s recovered a large part of a Soviet nuclear submarine, some of its nuclear missiles, and dead crewmen from deep in the Pacific. Would these slightly hot and hot accident scenarios be "catastrophic"

and cost billions to clean up? If we're just picking up a floating cocoon, the answer is "hardly." If we're using a deep ocean submersible, that cost must be determined.

But what if is smashes down on land somewhere, into an unyielding granite mountain? Would there be a breach of the cocoon and a release of radioactivity? My best guess is no, but this is easily answered in a research and development program because an enormous amount of data exists on the effects of things crashing into unyielding surfaces, from bunker busting bombs to earth penetrator weapons to airplanes crashing into nuclear power plants to a locomotive engine smashing into a spent fuel cask. I know that would leave the reader hanging for several years waiting for that program. It doesn't satisfy me so I know it won't you, but I can give an indication of what it might look like.

It probably would involve a demonstration – actually three demonstrations: Historical, theoretical, and practical. Historically, a wealth of data exists on the behavior of radioactivity at higher altitudes, particularly that from the atmospheric weapons testing period, and this would include the two exhaustive UN studies conducted in 1958. Much of this testing was done in the South Pacific, hence the relevance of such data to this launch profile. Then there is environmental data on the loss of nuclear submarines. Then there is Rover/NERVA data, as it had elaborate environmental precautions for all tests, with airplanes monitoring the effluents until they became impossible to detect. That and data from the National Oceanic and Atmospheric Administration would form a backdrop for the second phase: Theoretical or conceptual studies. With research and development tools now available, it is possible to simulate accident scenarios without any risks whatsoever. It's been done in the automotive and aircraft industries, and can be done here. So as engine and cocoon design progress, so can computer models that simulate the consequences of slight hot or cripple flight scenarios. It won't give hard and fast evidence, but it will show the demonstration phase where and how to go to give proof. That's the third part, a flight safety program that actually flies cocoons and mock engines (probably containing tungsten, not uranium, as they weigh about the same) and tests them under different accident scenarios, including ones in which accidents are deliberately planned and which might contain short-lived radioisotopes, such as those used in hospitals, to simulate a release. I cannot say how that would unfurl, but I am confident in our technological prowess to state unequivocally that I believe the risks of slightly hot or cripple scenarios can be reduced to the levels of the airline industry.

## VI. The Sixth Step: From 100,000-feet to LEO

The sixth step looks at the impact of a Re-core flying from 100,000-feet to 200-miles up. This is the success scenario and means the Re-core has started up in the upper reaches of the stratosphere (23,000-160,000-feet), fired through the mesosphere (30-50-miles) and reached LEO in the thermosphere (50-400-miles). What happens here?

All reactors have two types of radioactivity. First, fission products are the residues of the uranium atom splitting into two, into many other elements roughly half the atomic weight of uranium. These are radioactive, give off gamma rays and neutrons, and have different half-lives. In nuclear power plants, the fuel is contained in sealed rods to prevent the fission products from wandering around and contaminating the reactor. Rover/NERVA fuel development sought to prevent the loss of fission products, that's inherent in the 10-hour/60-recycle goal established in its final years, and it was well on its way to that.

I expect Re-core fuel development to emphasize retention of fission products, and I expect it to be very successful for the simple reason that it is easy, relatively speaking, while restarts are hard. How so? With restarts the fuel is going from near absolute zero to 2000°-3000+° C and then back again, repeating this over and over. Well, in the process, cracks form that do not seal up, so you get more and more cracks allowing uranium and fission products to escape. Think of bending a wire coat hanger repeatedly until it snaps in two. A Re-core would go only to high temperature once, minimizing the cracking process, thus minimizing the release of uranium and fission products. Nonetheless, some would just evaporate through the fuel so zero release is impossible. How much for a 10-15-minute flight is unknown now, but based on Rover/NERVA's record, I don't expect it to be much. The NRX-A6 ran for more than hour at full power, with no stops and starts, and the control drums, regulating neutrons in the core, barely moved. All this implies the fear of spewing radioactivity all over would be exaggerated, but this is an area for studies to clarify. It's also an area that fuel element development would affect dramatically; as better fuels appear, their release of fission products would decrease markedly.

Second are the gamma rays and neutrons, the most energetic and damaging radiation emitted during the fission process. These would be emitted as soon as the Re-core starts up, and would impact the stratosphere/mesosphere/thermosphere. However, these regions are already being bombarded intensely by cosmic and solar radiation, which contains gamma rays and neutrons (some of which strikes nitrogen molecules in the stratosphere to form radioactive carbon-14), and it fluctuates noticeably, particularly from solar flares that can pierce through these spheres and impact Earth. Indeed, a solar flare in 1989 knocked out power in Ontario, Canada for nine hours. So any added burden by a Re-core should be viewed as a minor, 15-minute max fluctuation. An analogy might be using a flashlight on a dark night. It lights the way as the person moves forward, but darkness envelops the path behind once again. So it would be with a Re-core, the gammas and neutrons "shining" on the stratosphere/mesosphere/thermosphere for a few minutes, but "darkness" returning swiftly at it flew to LEO. Any worry here seems quite exaggerated.

**VII. The Seventh Step: In LEO**

The seventh step means a successful injection into LEO. What are the hazards there? Two stand out. First, scientists might fear nuclear engines would cause a blurring problem, the radiation smudging the instruments on their satellites. This may or may not be a real problem and depends on too many factors to discuss here. Suffice it to say if it is a problem, the cocoon might mitigate it in the short-term (as it would dampen the radiation), while in the mid-term "garages" would probably be sent to LEO to house and perform maintenance on the Re-use engines returning from missions beyond. They would be shielded and prevent any blurring. Over the long-term, however, the introduction of heavy lifters fundamentally changes the nature of space science, allowing the economic construction and operation of a space science space station 20, 30, or 40 times larger than the International Space Station. Its sheer capability would dwarf all scientific satellites currently in orbit, and probably make them obsolete. Some scientists might object, fearing radiation blur, but most, I submit, would push hard for the newer station to come into being as quickly as possible. It would introduce a true golden age of space science.

Second, there might be a fear the Re-core would fall out of LEO and plummet back to Earth. Once an object attains a stable orbit, it remains there until its orbit degrades from the friction of colliding with the random air molecules present 200-miles up. This occurs over a long period of time, and a Re-core would not be there that long for this to happen. So this fear must be dismissed.

## VIII. The Eighth Step: Preparations to Return to Earth

After cooling down in LEO for a while, perhaps in the "garage," now it's time to prepare to return to Earth. In addition to locking the control drums in the "off" position, this might feature flooding the core with poison cement, a liquid glue that sets up in the holes in the fuel element, hardens like concrete, and prevents criticality during re-entry, making the core a solid, poisoned block, and preventing seawater from entering the reactor. For the core, it would not matter, as it is to be used once, then its uranium recycled and recovered. This return was never considered in Rover/NERVA; the closest to it was during the RIFT program that featured poison wires inserted in the core then pulled out as a Saturn rocket carried the reactor on a launch. No doubt many ways would be found to immobilize the core in the research and development process.

## IX. The Ninth Step: Re-entry

With preparations complete now it's time to step back to Earth. After a half century, the rigors of re-entry are well known and understood and an extensive body of know-how and experience exists to ensure it happens, without undue risk if the capsule contains astronauts and with reliability if a warhead is involved. Today few doubt that astronauts cannot return safely from Space, or that a nuclear warhead would fail to detonate as designed on its target, as fortunately has never happened. Most accidents occurred early on and were not unexpected; with added experience and new materials they were corrected. The cocoon would be designed and built from this wealth of knowledge, and at first it might resemble a warhead reentry vehicle, most likely spin-stabilized to distribute heat evenly and to act as a rifle bullet to descend on stable, straight, and accurate trajectory. A properly shaped re-entry cocoon should lower the speed to less than sonic (600mph), and the state of parachute development is well within its weight and speed specifications. Later models might resemble Orion/Apollo-like capsules, or like a lifting body that re-enters not on a ballistic trajectory but on a glide path to an isolated area, such as an island in the Pacific Ocean.

To suggest that the cocoon might disintegrate and scatter the radioactive reactor over the Earth is illogical. If space capsules, warhead re-entry vehicles, and the cocoon resemble one another technically and derive from the same body of knowledge and experience, all must succeed or all must fail. Logically, this says if A and B derive from X and succeed, then if C is similar to A and B and also derives from X, it must succeed too. To argue differently is to be arbitrary and selective. No one, to my knowledge, has ever opposed nuclear-tipped missiles because they feared the warhead would burn up and scatter its plutonium on re-entry; they opposed them cause they knew they would work as designed.

In the real world of practice and not logic, the cocoon would undergo a research and development effort that I spoke of earlier, the same as space capsules and warhead re-entry vehicles. This should not be difficult since the engine's weight is concentrated in a small volume, making it a larger version of a warhead; that is why the Titan's Mk-6 RV is so intriguing. The concentration of weight makes it easier to achieve the directional balance and stability so critical to attain as an object re-enters the Earth's atmosphere. At the end of this development, I believe the cocoon would demonstrate it could return intact under all plausible accident scenarios without releasing any radioactivity.

## X. The Tenth Step: Recovery

Now I have reached the last step in the walk-through. The cocoon has landed in the ocean somewhere, and since the cocoon is buoyant, recovery is easy. Just bring a ship next to it, lower a cable from a crane, and lift it into a shielded cask on deck. This would not be a lengthy or complicated process so any radiation exposures to the crew would be minimal; the navy and nuclear industry have more than half a century's experience in such operations, that it would be at sea on a pitching deck adds just a new wrinkle. Any worry of a radioactive cargo on a ship must be firmly dismissed. As you read this, the nuclear navies of many countries and cargo ships containing irradiated fuels are sailing the high seas and have been doing so for decades.

## Summary

Thus I end my ten-step walk-through. In each step, I tried to highlight key problems and indicate how they might be solved or reduced to manageable levels of risk, ones most of us accept for many conveniences of modern life, such as air travel. I hope the reader understands they are not intractable, but can be solved in a straightforward manner by our scientists, engineers, and technicians. Others may raise the specter of catastrophic accidents. Here I say hold on. Don't believe me, but don't believe them either. Withhold your judgment. Adopt the motto of the state of Missouri: "Show me." Let our technical people work on this never-studied topic, let it crystallize, and then let it be made available to you. Then let our range safety experts weigh in; their integrity, knowledge and judgment must be heard. Then in public forums let those who support "show you" their data, those who oppose to "show you" theirs, and range safety experts "show you" everything they know. Demand evidence, not emotion. When given that, you will likely have two distinct and different viewpoints as well as those of range safety experts, giving you the opportunity to judge. I believe when this evidence is presented, it will demonstrate overwhelmingly any "risk" will be equal to that of the airline industry initially and with experience, it will become better than it. That leaves you to measure the "risks" in your mind against the "gains" not only in your wallet but also in your life as a new epoch in human history has been introduced. You have personal access to Space, as well as rights over the solar system.

# Index